W
THEY
HID THE
FIRE

INTERSECTIONS: ENVIRONMENT, SCIENCE, TECHNOLOGY
SARAH ELKIND AND FINN ARNE JORGENSEN, EDITORS

WHEN THEY HID THE FIRE

A History of Electricity
and Invisible Energy in America

DANIEL FRENCH

UNIVERSITY OF PITTSBURGH PRESS

Published by the University of Pittsburgh Press, Pittsburgh, Pa., 15260
Copyright © 2017, University of Pittsburgh Press
All rights reserved
Manufactured in the United States of America
Printed on acid-free paper
10 9 8 7 6 5 4 3 2 1

Cataloging-in-Publication data is available from the Library of Congress

ISBN 13: 978-0-8229-6425-4
ISBN 10: 0-8229-6425-2

Cover design by Laurence Nozik

CONTENTS

PREFACE VII

INTRODUCTION 3

1. ENGLISH ROOTS, UTOPIA FOUND AND LOST 20

2. THE ENERGY REVOLUTION AND THE ASCENDANCY OF COAL 31

3. THE CONUNDRUM OF SMOKE AND VISIBLE ENERGY 46

4. TECHNOLOGY AND ENERGY IN THE ABSTRACT 60

5. OF FLUIDS, FIELDS, AND WIZARDS 82

6. ENERGY, UTOPIA, AND THE AMERICAN MIND 102

7. TURBINES, COAL, AND CONVENIENCE 122

CONCLUSION 142

NOTES 159

REFERENCES 201

INDEX 231

PREFACE

The morning of August 3, 2014, began as a typical day. I woke up and began my routine of filling the kettle with water and putting it on the stove. As I waited for my water to boil, I checked my messages and noticed a text from a friend that said, "Don't drink the tap water and turn on the news." I turned on the television and learned that there was a ban on our water as it had been found to be unsafe to drink. In an instant, half a million people in northwest Ohio and southeast Michigan were without water. A toxic algae bloom on the western end of Lake Erie had settled over the Toledo, Ohio, water plant intake crib and suddenly a major Great Lakes city had no usable water. As I write this in March 2016, just one hundred miles north of my location on the University of Toledo campus, our neighbors in Flint, Michigan, are in the midst of their own water crisis. A switch in their water source caused lead to leach out of century-old pipes, rendering the water unusable. In both cases, systems that were taken for granted became the center of attention. It was rare that one would think about the water supply in either city considering their locations near the largest sources of freshwater in the world.

It may seem strange to begin a book about electricity with a story about water, but the water supply shares a common element with the electrical grid. They are both largely invisible technical systems that we tend not to think about. In the case of our water crisis, there was a sudden and finite interruption that exposed vulnerabilities in the system. There are well-publicized vulnerabilities in our electrical grid as well, but the subject of this book is not about immediate interruptions or the lights going off. My area of interest is in how the invisible nature of our electrical infrastructure contributed to attitudes about energy consumption and the environmental consequences of that consumption over time. Like water, electricity is a commodity we encounter every day, but unlike water, it is in itself invisible, and we rarely see the ultimate sources of electrical generation.

As a secondary energy source, electricity has always been somewhat of an abstraction. Electricity is a secondary source because it requires a primary energy source in order to be generated, such as a photovoltaic cell or mechanical action derived from steam or wind. As such, it is once removed from its ultimate source of generation and represents an energy source that takes on a new identity, further removed from the consciousness of those who consume it. I have taken an informal poll with my undergraduate students every semester for the past five years with one question: What is the ultimate power source for your smartphone? Most students quickly answer that it is the battery. I then ask again, what is the *primary* power source? What ultimately charges your wireless device's battery? Eventually we get to the answer that it is electricity, and where I teach, many go on to identify that the ultimate source of electricity is nuclear power, likely because we have two nuclear plants within forty miles of our campus. A few others answer correctly by saying that the primary source is coal. It is a wake-up call for some to realize that our cell phones, plug-in cars, laptops, and all other ubiquitous devices in large part run on coal. Most of us do not consider the fact that nineteenth-century steam-turbine technology powers our smartphones and other modern electronic devices. We've become consciously removed from the energy infrastructure, and we have come to view electricity as an autonomous, stand-alone energy source. In short, on a daily basis, we have forgotten where our energy comes from and the consequences of our energy consumption.

This did not happen overnight. We've had a long love affair with electricity. When electrification was adopted in the United States at the end of the nineteenth century, it was an improvement over the smoky coal, wood, and oil fires that heated and illuminated homes and businesses. Coal was still involved, but it was removed from view and its power was silently transported by way of innocuous wires. As power generation plants moved farther away and the visible components of coal smoke were no longer seen, it was easy for society to forget that there might be consequences to our insatiable thirst for electricity. Many of us thought that our electricity came from famously "clean" and well-publicized hydroelectric plants such as Niagara Falls or the Hoover Dam. As electrical generation facilities moved farther from view and the infrastructure became part of the landscape, we simply forgot where our energy came from and the harm it was causing to the environment.

As a historian of technology, I wanted to know more about when and how this all happened. Did electrification affect how society would view and consume energy? How did the invisible nature of the power grid shape perceptions of energy consumption and the environment? How did cultural messages shape society's views on electricity and energy in general? These and many other questions informed my research for this work. I departed from many existing studies by examining how technological change—in this case, electrification, shaped perceptions about energy and the environment. A useful analogy may be that of Benedict Anderson's concept of "Imagined Communities" only in the proposed study it is "Imagined Environments" that I am concerned with.

My research seeks to determine if a cultural discourse of environmental inconsequentiality emerged as Americans transitioned from salient energy sources that involved direct contact with fire to electricity which was clean, silent, and invisible at the point of consumption. I was also interested in an American cultural ideology I identified as "energy exceptionalism," and whether or not unlimited energy usage has always been assumed to be a right in the United States.

As with any research project, more questions are raised than are answered. Are those who remain climate change deniers informed by long-standing cultural messages that equate the United States with unlimited and inconsequential energy? How have more visible infrastructures such as solar arrays and wind turbines informed public opinions on energy? Has energy consciousness and literacy changed as climate change evidence becomes irrefutable? Much more work needs to be done as we are facing a critical juncture in coming to terms with our nonchalant attitudes toward energy consumption, and it is my hope that this work will contribute to increasing energy literacy.

As I have quickly learned, there is a lot more to authoring a book than one might assume. Just when you think that the long nights of research and writing are through, you learn that publication presents a new set of challenges. Thankfully, the staff at the University of Pittsburgh Press has been patiently there for me. From my editor, Sandy Crooms, to copy editor Amberle Sherman, everyone I have worked with at Pittsburgh has been professional and helpful, and I cannot thank them enough.

Many others have contributed to the shaping of this work and my intellectual growth throughout the process of this research. This project would have never gotten off the ground without Dr. Diane Britton, who encouraged me from start to finish at a time when few others expressed interest in the topic. Through our shared affinity for the outdoors and blue skies, we found common ground in tracing the roots of society's lack of urgency regarding the issue of climate change. There are numerous other colleagues who helped more than they know. Dr. Kim Nielsen at the University of Toledo who has always had a smile when needed and sound advice, Dr. Peter Linebaugh, who provided valuable input throughout the work, and Dr. Daryl Moorhead who brought a much-needed and appreciated perspective of an environmental scientist and ecologist to the project.

I am also grateful to my coworkers at Mercy College of Ohio. I am fortunate to be surrounded by a group of individuals committed to intellectual inquiry who have provided much-needed social and emotional support. In addition to my colleagues in academia, I'd like to thank the many friends who have politely listened to me harp on about the work and helped divert my attention from the rigors of publication. To my friends in the Sunday Ann Arbor acoustic jam group and in Darrell Scott's Songfood Workshop in Nashville, thank you for putting up with my ongoing attempts at music and keeping me relaxed and distracted from the stresses of writing and research; your friendships have meant the world to me. I'd also like to thank both Darrell Scott and Tim O'Brien for the insight that their song "Keep Your Dirty Lights On" has brought to me throughout my research. I have found that a few hundred well-written lyrics often express complicated ideas more effectively than volumes of academic writing ever could. Music has been a much-needed diversion for me throughout my research, and many friends in the Toledo music community have kept me entertained and relatively sane while I pondered big thoughts in various dark venues throughout the area.

Last but by no means least, thank you and much love to my spouse Anne and my son Joe. Your patience and encouragement are appreciated more than any words can express. We turned long family trips traveling from Ohio to the coal pits in Wyoming and fracking sites in North Dakota into valuable research vacations that I will never forget. Without the support of my family, this book would have never been completed. Anne was a sounding board throughout my education and research who always kept saying, "Go finish the book." Now I can say that it is done.

WHEN THEY HID THE FIRE

INTRODUCTION

On October 2, 1927, the *New York Times* featured a bold headline: "The Electric Age: A New Utopia." Explaining history in terms of energy epochs, the author declared that the United States was facing the most remarkable transformation in its history: "It began with the steam engine and the first industrial revolution. It is now closed by the electric superpower system and the new industrial revolution."[1] In closing the age of steam, the author reflects an ideology of energy exceptionalism that positioned electricity as an autonomous source of power that was modern, unlimited, and a clean replacement for the old-world technologies. Absent from the article was any mention of coal or steam, even though nearly all of the electricity generated in the country at the time was produced by burning coal.

Fifty years earlier, when the Centennial Exhibition in Philadelphia introduced the first practical applications of electricity to the country, they were mere curiosities. The 1876 world's fair featured the majestic Corliss steam engine, which captivated many of those who attended, as well as a host of other new technologies, including the telephone, the improved telegraph, and several arc lighting systems. While the Corliss ran on steam transported to the engine via underground pipelines, the telephone, telegraph, and arc lights received power through unobtrusive strands of wire.

Electrical power was an entirely new energy paradigm, unlike anything that came before it. At the point of consumption, electricity created light and heat, or provided a backbone of current on which communications could travel. It was at once miraculous and mysterious. Electricity was antithetical to open flames and hissing steam: it was intangible, it was silent, and it was invisible. Whereas the utilization of energy from fire or steam was only possible close to the point of combustion, electricity extended the range that useful power could travel. By inserting physical space between smoky fires, pressurized boilers, and usable power, the small dynamos present at the fair gave life to clean, silent electrons that became surreptitious energy agents for dirty, burning coal. Embraced by the American public as a modern and progressive power source, electricity began to replace the fires of gaslights and was poised to power a variety of contrivances in homes and businesses. In a technological sleight of hand, coal was converted to electricity in a process removed from the view of the rising American consumer class.

Writing in 1906, historian Henry Adams recognized that electricity was "but an ingenious channel for conveying somewhere the heat latent in a few tons of coal hidden in a dirty engine house kept carefully out of sight."[2] Adams's comment identifies a key feature of how energy would become conceptualized in the American mind—as burning coal and central power plants moved farther from sight, society would begin to embrace electricity as an environmentally inconsequential source of energy. With his contention, Adams recognized that electricity represented invisibility and a loss of direct contact with fire and fuel, a technology that transformed dirty, old-world energy into something modern and clean, and subsequently became portrayed as autonomous.

Although electrification did not eliminate all industrial fires nor solve all of the problems of industrial smoke, it did begin to eliminate the flames that society encountered on an up-close and regular basis. As it became widely adopted in the early 1900s, electricity eliminated the salience of energy use by removing the need to handle fuel, tend to flames, or experience the detritus of smoke or soot. Clean and smoke-free, electricity fit well with deeply engrained visions of a bucolic and pastoral America. At the same time, the always-on, unlimited nature of electric power altered the awareness of consumption. Prior to electrification, coal bins, woodpiles, and lantern-cans provided a visual—and physical—indication of the amount of energy used. The electric switch did not. These two qualities of clean energy at the source and limitless on-demand power fit well with long-held American attitudes about consumption and consequences. The delivery of electric energy via wire, along with other forgotten infrastructures such as water distribution systems, constituted what historian Martin V. Melosi has described as "hidden functions."[3] While electrification did not eliminate industrial smoke in the Progressive Era, it transformed domestic energy use into a hidden function that became innocuous and unlimited in the American mind.

Unlimited and inconsequential energy fit well within an ideology of energy exceptionalism, a condition that had existed in American culture since the first European colonists arrived. In North America, where fuel sources seemed inexhaustible and the environment infallible, American attitudes toward energy use developed around ideas of inconsequential consumption. As industrialization and fire-based energy began to threaten exceptionalist visions of America's pristine nature in the second half of the nineteenth century, a new urban middle class was ripe for progressive solutions. The subsequent development and adoption of electrification began to reframe American attitudes toward energy use. As an imagined alternative to fire, steam, and coal, electricity came to be seen by consumers as an energy panacea. From its early commercialization and moving forward, advancements in technology began to disassociate electricity from the coal and steam that produced it. Electricity became an abstract form of power as technological, cultural, and social factors combined to assign new social meanings to energy use. As technological advancements allowed for increased physical distance between power generation and power consumption, the commodity of electricity became an independent actor, consciously detached from the infrastructure of production. As the spatial dynamics of energy production and transmission changed, cultural factors led the public to view electricity as mysterious, utopian, and an alternative to the proximal fire-based energy sources of the past. With the adoption of electricity occurring simultaneously with the trends of Progressivism and consumerism, power companies promoted the use of electricity while energy infrastructures became less visible. As electricity became disassociated from coal in the minds of Americans, an ideology of energy exceptionalism reformed around renewed beliefs of inconsequential consumption.

Electricity has been perceived by American society as a modern, unlimited, and clean form of energy since it came into practical use at the end of the nineteenth century. As power generation plants moved out of city centers and dark particulates were scrubbed from smokestacks, consumers had no reason to believe that electrical power was anything other than a clean and progressive energy source. Removed from their consciousness, the burning coal that ultimately powered their world on the other side of the outlet became forgotten by Americans. For nearly a century, the billions of tons of coal that powered the electroconsumer culture in the United States was, as Adams said in 1906, hidden in the "dirty little engine house" that was the electrical generation infrastructure.

Human beings tend to respond to threats that are easy to picture, and electrification allowed for the hiding of the deleterious consequences of its production. Even after the effects of nearly a century of coal-fired electricity generation began to surface in the press in the early 1970s with the discovery of acid rain in the northeast United States, the association of coal with electrification never fully materialized in the mind of the American consumer. Recent studies suggest that a majority of Americans are still not aware of the origins of their electricity and the role that coal plays in its generation.[4] The invisible nature of the electrical production and distribution infrastructure has led to a "blue-sky" mentality—if the sky is blue and looks clear then everything must be okay. As this manuscript goes to press, technological and energy abstraction has continued to accelerate. While Americans' energy consciousness and literacy may now be changing in light of overwhelming evidence of global climate change, the role that coal-fired electricity plays in our lives is still largely an abstraction. Few of us think about the nineteenth-century technology of coal and steam that produces much of the electricity required for our ubiquitous Internet, smartphones, or wireless tablet computers. Electric automobiles with names such as Leaf, Volt, and Tesla are seen as technologies that are freeing us from the bonds of fossil fuels, yet in many locations they too largely depend upon coal- and steam-derived electricity for their locomotion. Few think past the marketing hyperbole or are cognizant of the seven hundred million tons of coal still burned per year to power our illusionary modernity. We are as a society better informed now than we were in the past, yet nonconscious consumption as it applies to energy usage remains problematic. While this work is limited to showing the historical roots of American energy exceptionalism as it relates to electrification, it also intends to shed light on our present condition. Today's remaining skepticism over climate change and insensibility to the consequences of energy use is directly informed by the past, and in this story we can value history as a mechanism that drives awareness of the consequences of our deep-seated and ongoing cultural behaviors of consumption.

If this work advances our understanding of how Americans perceive their technological world, it is by standing on the shoulders of giants that have come before. Past studies have traced how Americans reconciled the ascendancy of technology with the sublimity of nature. This work attempts to expand the lens of inquiry into less-visible technological systems in general and energy systems in particular. Works that have influenced this book include Leo Marx's *The Machine in the Garden*, from 1964, which was a breakthrough in merging intellectual history, the history of technology, and culture. Focusing on the nineteenth century, Marx examines how Americans came to resolve ideas about the environment and progressive views of technological and scientific advancement. Marx's work is notable in that it shows how American writers and artists began to merge technology into the environment, creating a "middle landscape" between an unspoiled primitivism and a technologically advanced progressivism.[5] By illustrating how industrialization became merged with pastoralism in cultural artifacts, Marx shows how technology became reconciled as part of nature. In his deconstruction of Henry David Thoreau's *Walden*, Marx reveals harmony between the organic and the inorganic, in which "the hills in the background and the trees of the middle distance gently envelop the industrial buildings and artifacts. No sharp lines set off the man-made from the natural terrain."[6] Concluding that Americans have embraced an ideology that technology, "the machine," could coexist with a pure environment, "the garden," Marx follows the roots of a pastoral ideal back to the time of Jefferson. While the country grew as an industrial powerhouse, Marx argues, a "technological sublime" ideology was promulgated that positioned technology as a panacea, and this ideal is often reflected in the American historical discourse. Although Marx is making an argument that views industrialization as destructive and ultimately in conflict with nature, his epistemological basis that mechanization and technological advancement led to an American rhetoric of the technological sublime is a framework that can serve well as a starting point for examining cultural ideas related to energy usage. From atomic power to the internet, scholars have often speculated over the promise of new technologies as a positive force of change—with utopian dreams serving as blinders to potential social or environmental effects. The story of electrification runs along very similar lines.

A number of other notable authors have traced the effects of technology and industrialization on the environment. These works not only investigate environmental exploitation and pollution but also are useful in following the path of environmental reform and consciousness in the United States and relate well to society's incognizant use of energy. David Stradling's *Smokestacks and Progressives: Environmentalists, Engineers, and Air Quality in America, 1881–1951* examines how coal-related smoke abatement movements in the United States formed in response to the growing coal smoke problems that arose between 1881 and 1951. Stradling's book, which is as much a social and technology history as it is an environmental work, examines shifts that took place in the makeup of reform leaders as well as in the ways reformers confronted the smoke problem. In addition to discussing the activities of reformers, Stradling's book offers insight into how a new rising class of engineers began to displace laypeople as the stewards of airborne emissions.[7]

While Stradling examines the nascent environmental issues that confronted bituminous cities such as Cincinnati, Chicago, Cleveland, and Pittsburgh, William Cronon's *Nature's Metropolis: Chicago and the Great West* explores the effect that urban growth had on the surrounding countryside.[8] Cronon addresses how space between producer and consumer became irrelevant as Chicago's thirst for commodities expanded outward. In this work, he explores two highly relevant themes: spatial dynamics and commodification. Although Cronon's study addresses people's relationships with space in the context of the expansion of railroads, his broader point is that as the distance between production and consumption of goods grew, natural products lost their identity. Cronon uses goods such as meat to make his point, stating that the distance between "the meat market and the animal in whose flesh it dealt seemed civilizing."[9] While he does not address energy in his work, Cronon's examples relate to the distance between the coal mine, the dynamo, and power consumption. Just as the railroad was the facilitator of space in Cronon's study, in an examination of energy usage the wire is the connection as coal is transitioned into its more "civilized" form of electricity.

As a collection of essays that touch upon the themes of both Stradling and Cronon, Martin Melosi's *Effluent America: Cities, Industry, Energy, and the Environment* also examines the history of urban pollution and urban environmental reform.[10] *Effluent America* frames a range of issues in urban environmental history, including the relationship between industrialization and pollution as well as a study of the effects of urban growth. Melosi's section on coal and smoke in the late nineteenth century is particularly relevant because it reveals how smoke was associated with a "degraded" society. Quoting from Lewis Mumford's *The City in History*, Melosi relates to Mumford's "Paleotechnic Paradise: Coketown," which in turn borrows from Charles Dickens's *Hard Times* the notion that "Coketown specializes in producing dull boys."[11] Only when the cleanliness associated with electrification is seen against this backdrop of evil smoke can one gain a clearer picture of the juxtaposition of electricity, which is a core theme of the present work.[12]

While environmental studies and intellectual histories examine the implications of anthropocentrism and how Americans attempted to reconcile and justify technology, histories of technology explore how infrastructures were built in response to social acceptance. The evolution of electrical networks that began in the latter half of the nineteenth century occurred with great public fanfare and occasional controversy, yet as the technology and transmission networks matured, social acceptance was rapid and widespread. The two most widely recognized works on the study of electrification are those of Thomas Parke Hughes and David E. Nye. Hughes's book, *Networks of Power: Electrification in Western Society, 1880–1930*, is an all-encompassing study of electrical systems in the United States, Great Britain, and Germany. Hughes is mainly concerned with examining electrification as a system, and as such, concentrates on the design and construction of electrical grids and interconnectivity.[13] In documenting how long-distance power transfer developed in the western United States, Hughes shows how this development was a gateway for the spatial dynamic that allowed for increased distance between electrical generation and consumption.[14] Hughes also dedicates an entire chapter to the development of electrification in Chicago, a topic of great importance in this study. In Chicago, under the direction of Samuel Insull and the Chicago Edison Company, the first urban system that merged smaller power networks into a regional grid formed. Thus, the generation of power became further removed from the central city much earlier compared to other urban areas.[15]

David Nye's *Electrifying America: Social Meanings of A New Technology, 1880–1940* is a technological and social history of electrification that traces not only the adoption of systems but also how Americans confronted the new energy source.[16] Nye's work is broad-ranging, covering a variety of areas from the development of the grid to the White City of the Chicago World's Fair in 1893. Of all the existing literature, *Electrifying America* most closely pursues the social meaning of electrification expressed in utopian ideas as well as the evocation of electric landscapes in art and literature.

Nye is particularly insightful in examining how the public confronted the new technology of electricity. In one key section, he identifies a "small technical elite that viewed [electrification] as an instrument for rationality and social reform."[17] This group runs congruent to the rising class of professionals identified by Robert Wiebe who were confident that engineered solutions would trump old-world problems as represented by coal and smoke.[18] Faith in engineers allayed concerns about the environment, and Nye's contention that the emergence of an engineering "elite" in the late 1800s signaled a cultural shift from confidence in traditional high culture to confidence in scientific knowledge is significant. This view runs parallel to David Noble's idea that "the electrical and chemical industries form the vanguard of modern technology in America."[19] Nye's work identifies many of the key factors in electricity's social history; as such, his work serves as a launching point into deeper inquiries into how the technology shaped Americans' attitudes.

Whereas both Hughes and Nye treat of the social adoption of electrification on a macro level, Maury Klein's approach in *The Power Makers* considers the incremental development of energy technologies and the transition from steam to electricity. Klein's work is valuable in that it documents the long, slow process in the discovery of the "mysterious ether" that was electricity.[20] Here, the "terrifying force" was eventually harnessed, but not without a long pedigree of being associated with "lightning and divinity."[21] While Klein's study concentrates more on technological inventions than perceptions, his history does illustrate how electricity was viewed as a more advanced energy source than coal. In addition to discussing technological development, Klein also provides a well-rounded overview of the key personnel in the development of electrical systems. While the stories of Edison, Tesla, and Westinghouse are well known, Klein's research on Insull begins to address an important point in how society began to see electricity as environmentally inconsequential.

Insull was a pioneer in consolidating small power generation stations and creating a model for wide area transmission networks, a critical point in the separation of power generation from consumption. In his investigation of Insull's archives, Klein reveals evidence that reducing smoke pollution from the center of Chicago was on Insull's mind as he installed new turbine dynamos in the early 1900s.[22] While Klein's work as well as that of Forrest McDonald documents Insull's impact on the development of the infrastructure of electrical transmission, the broader implications of Insull's model are not addressed.[23] By offering inexpensive power, Insull promoted an "always on" mentality that encouraged energy consumption. At the same time, the regional networks that Insull pioneered, which were powered by large-capacity steam turbines, locked in coal as the primary fuel for electrical generation in the United States.

A number of other works are significant in any study of perceptions of energy and energy transitions in the United States. *Routes of Power: Energy and Modern America* by Christopher F. Jones is masterful in tracing the roots of fossil fuel dependence in the United States.[24] Jones's work goes far to explain how society's desire for inexpensive energy led to our current predicament of nonrenewable energy dependence. Vaclav Smil's *Energy Transitions: History, Requirements, Prospects* is a valuable study that provides a scientific and technical analysis of energy usage while documenting critical turning points in the nation's energy history.[25] Last, but by no means least, is the work of Richard F. Hirsh and Benjamin K. Sovacool, specifically their paper titled "Wind Turbines and Invisible Technology: Unarticulated Reasons for Local Opposition to Wind Energy."[26] Hirsh and Sovacool's work should be required reading for anyone interested in how Americans view energy use. By bringing a perspective of the current state of nonsalience surrounding the nation's electrical infrastructure, Hirsh and Sovacool provide an up-to-date ending to the history contained in this work.

While many of these works investigate how technologies and energy sources have developed and been adopted over time, the present study examines how American attitudes about energy evolved in the context of electrification. The first chapter explores the cultural roots of energy exceptionalism in the United States. Since the first Europeans arrived in North America, the attitudes they developed toward energy were shaped by the notion of abundance. Coming from an England that had been deforested due to the prolific burning of wood, the first settlers saw the New World in terms of an energy bounty with no shortage of consumable energy. Along with the importance early colonizers placed on ample fuel, narratives that they left behind reveal an appreciation for the clear air of a utopialike unspoiled continent. The contradiction within these two values—a desire to exploit resources of energy within the framework of a pristine environment—sets up the foundation of energy exceptionalism. As the nation evolved, these incompatible values remained in place; American society became a voracious consumer of energy yet continued to imagine the country in terms of pastoralism and environmental tolerance.

Chapter 2 shows how the contradiction between energy-intensive economic development and environmental inconsequentiality evolved. In the first two hundred years of European occupation, North America began a transition from a pristine natural landscape to an industrialized society similar to what was left behind in England. Although pristine pastoralism was a state that many in the early United States wished for, industrial capitalism became a growing force. Thomas Jefferson's vision for the country was a low-energy society of agriculture and yeoman farmers, where the bounty of the land could offset the need for industrialization and urbanization. Jefferson's ideal of the country as a rural space remained embodied within an ethos that scholar Richard Hofstadter identifies as the "agrarian myth," and Leo Marx refers to simply as "the garden."[27] Despite Jefferson's vision of what the country should be, the reality was something entirely different. In a quest for national wealth and power, men such as Alexander Hamilton, Tench Coxe, and even Jefferson eventually accelerated domestic manufacturing and urbanization.

The nation's first textile mills in New England fit within the ideological framework of a pure environment. Water-powered and smoke-free, the mills were seen by many observers as exceptional and superior to the Dickensian factories of the Old World. Despite the clean energy incorporated in these commercial utopias, they were short-lived. An unquenchable demand for profit and goods outpaced the capabilities of water power, and by the middle of the nineteenth century the coal-fired steam engine was becoming widely adopted in the United States. By the end of the Civil War, the nation had set a course to be a society dependent upon carboniferous fossil fuels.

While energy exceptionalism as an ideology began with the first European colonists and remained a part of the American ethos, the inherent conflict between pyrotechnologies and an unspoiled environment began to surface in the nineteenth century. Following chapter 2 tracing the ascendancy of coal and its inconvenient cognate of smoke, chapter 3 explores urban American society's response. Pressed between conflicting values of production, consumption, progress, and social health, urban reformers and industrial capitalists were at odds over the effects of high energy consumption. Industrial boosters and those who stood to profit took the position that smoke was beneficial, while middle-class urban reformers began to equate smoke with social degradation and ill health. The conundrum for a rapidly urbanizing population in the cities of the Midwest and East was that of a modern society dependent upon an ancient form of fire-based energy. As the imagined ideal of a pristine and forgiving environment became obscured, inventions and efforts to control or hide the smoke represented an impulse to restore a sense of pastoralism.

Chapter 4 demonstrates how technological solutions and the emergence of technical systems and networks began to instill confidence that reconciliation between the conflicting values of consumption and a pristine country was possible. Twenty percent of the nation's population attended the Philadelphia World's Fair in 1876 and witnessed the emerging technology of electricity. In a society that was illuminated, heated, and powered by proximal fire, the electrical technologies on display at the fair debuted a new paradigm of energy abstraction. Suddenly, electricity altered the spatial dynamics of energy production and consumption by transforming the heat energy bound up in coal into a new form of power and transmitting it through innocuous strands of wire. Whereas steam power and all other sources of energy necessary for the production of light, heat, and mechanical motion required proximal flame and visible networks of distribution, electricity possessed the ability to distance flame from consumable energy. In an exploration of the 1876 fair, chapter 4 uncovers not only technological shifts but also how influential social leaders began to define electricity as an energy panacea that could transform society. While the physical separation between fire, steam, and electricity represented an instance of the technological abstraction of energy, the rhetorical portrayal of electricity as an autonomous antithesis of coal marked the beginnings of a cultural abstraction.

Chapter 5 examines the roots of energy abstraction in the context of electrification in greater detail. Invisible and intangible, electricity was easily detached from nature and well aligned with the ideology of exceptionalism. The chapter traces the development of electricity from the early theoretical work of Benjamin Franklin to the laboratories of continental Europe and England. Conceptualizing electricity proved difficult, as even those well schooled in the sciences could not entirely grasp what it was. Franklin saw electricity within the metaphor of fluid, and scientists such as Alessandro Volta and Hans Christian Ørsted reconceptualized electricity's properties when the fluid metaphor proved inadequate. Although the advancements of theoreticians such as Michael Faraday and Joseph Henry led to the practical application of electrical power, the difficulty in understanding the nature of electricity remained. Electricity was unearthly and ethereal as it was rendered invisible and transmitted through wire. From the inventions of men such as Samuel Morse to the first lighting systems of Charles Brush, the public was captivated by the mysterious nature of electricity. As power generation moved farther away from power consumption with the deployment of the first centralized and regional generation stations, the energy supply chain became less connected and more abstract.

Following the previous chapter's demonstration of how technological abstraction began to render electricity as an invisible energy source, chapter 6 explores the process of cultural abstraction. Electricity as a new energy source became not only physically detached from the coal and steam that ultimately created it but consciously detached as well. A core tenet of energy exceptionalism is the perceived inconsequentiality of energy production and consumption, which began to occur when both the physical and cultural manifestations of the electrical delivery chain became hidden. As the physical infrastructure was being obscured by distance, electricity became culturally constructed as a utopian power source. By 1882, the perception of electricity had changed from a technology that was a curiosity to one that was a utopian source of energy. Attitudinal shifts were due not only to technical advances such a long-distance power transmission but also from cultural cues. At the Chicago Columbian Exposition in 1893, organizers of the world's fair created an idealized electric dreamscape known as the White City. To the twenty million people who attended the fair, electricity became synonymous with American progress: it was clean and capable of almost anything. At the same time, plans for harnessing the power of Niagara Falls were announced, leading to news accounts that unlimited and inexpensive energy was soon at hand. As the White City and the promise of Niagara gave hope that a smoke-free environment was possible, the period's literature promoted and reinforced visions of an electrical energy panacea. In the popular utopian novels of the late nineteenth century, electricity was the featured technology that rescued society from the smoky haze of oppression. The chapter ends by showing how the White City, Niagara, and period literature promoted a mind-set that disassociated electricity from environmental consequences.

Although by the early twentieth century many segments of American society believed that electricity was a utopian and inconsequential energy source, the reality was something else entirely. Chapter 7 examines the formation of the modern electrical infrastructure and investigates how unlimited electricity consumption was encouraged by the rising trends of consumerism and inexpensive power. In addition, writings and media accounts of the time show how Americans had come to define electricity as a stand-alone energy source completely disassociated from coal. While the possibilities of maintaining a society of limitless power consumption within a pristine environment wavered in the fire-powered years of the Gilded Age, the exceptionalist legacy was restored to the American mind after the turn of the century. The chapter shows how the adoption of the steam turbine as the nation's primary source of power generation further obscured the nation's electrical infrastructure and inextricably tied the nation's electricity production to coal. First utilized on a large scale by Samuel Insull and the Chicago Edison company in 1903, steam turbine technology resulted in higher-capacity generation plants that led to fewer power stations farther removed from the populations they served. Consequently, the electrical infrastructure became less visible to the consuming public. In the process, the coal that fired the plants was obscured from the sight and consciousness of American consumers. As the nascent electrical grid became more concealed, electric companies promised a utopia of convenience. By encouraging the purchases of electrical contrivances power companies promoted electricity as an energy source that freed the housewife from toil and encouraged men and boys to build and experiment with electrical tools and toys.

As the adoption of electrical conveniences and the consumption of electricity per household expanded, the abstraction of energy grew as well. Electricity became defined as a clean energy source in the early twentieth century and it became clear that the link between electricity, coal, fire, and smoke had been broken in the mind of society. While the American public celebrated clean electricity, annual coal usage in the United States continued to grow. In 1927, when the *New York Times* proclaimed that the "electric age" had arrived and the age of steam had passed, the country was poised to consume more coal than ever before. At the same time, Americans became steeped in a culture that encouraged unlimited power consumption that came with no environmental consequences. As the electrification of the country expanded, the coexisting notions of clean and unlimited energy intensified, mirroring the ideology of energy exceptionalism first espoused by European colonists upon landing in the New World.

For historians, research is always a double-edged sword. At once we are thrilled to uncover intriguing stories that were unexpected finds, but frustrated we cannot follow every thread we uncover. In this history of energy and electrification in the United States, many possibilities for alternatives to coal-based energy appeared along the way, and the parallels to the present go far to reveal the roots of our current embedded energy infrastructure. As early as 1833, German immigrant John Adolphus Etzler surmised that there were natural forces such as the sun and wind that could be harnessed to drive the future. Etzler appreciated the progress that steam-powered mechanization could bring to society, yet he felt that the coal interests would have too much power over the masses. Although Etzler did obtain patents for his early wind technologies, he found no support for his pursuits and was ultimately seen as an eccentric.

In 1853, John Ericsson, who would go on to invent the iconic Civil War ironclad the USS *Monitor*, built and sailed a two-thousand-ton "caloric" ship, the *Ericsson*, between New York and Washington solely on the power of the sun. Despite this accomplishment, the ship was deemed as too slow for commercial purposes.[28] A little more than two decades later, Ericsson was shut out of displaying his caloric engine at the Philadelphia Centennial Exhibition in 1876, which was largely sponsored by coal interests.

In 1888, Charles Brush, an early pioneer of alternating current electrical lighting systems, built a wind-powered dynamo to supply electrical power to his Cleveland home. Despite his success with the project, Brush's technologies were not pursued, as coal-derived electrical generation was already becoming a well-entrenched technology. While Etzler, Ericsson, and Brush may be three of the more well-known alternate energy enthusiasts, archival research has exposed others who were pursuing a noncoal infrastructure for the generation of electricity.

In a delightfully accidental find deep in the archives at the University of Wyoming, correspondence between a Laramie rancher, Frank Bosler, and the General Electric Company, reveal another attempt in the pursuit of electrical generation without coal, utilizing wind power. The letters from 1913 between Bosler and General Electric—one of the major manufacturers of steam-turbine electrical generation components—demonstrate General Electric's commitment to coal-turbine technology.[29] The company's "wind will never work" philosophy both ensured that the United States would become a voracious consumer of coal and set back any viable pursuit of renewable energy at the same time. As a backstory in this work, the history of renewable energy in the United States is an area ripe for further inquiry. As alternate energy sources show up chronologically in this work, they are perhaps an indicator of how a noncoal society could have evolved, yet as history is not always a place for counterfactual analysis, the reader will have to decide what might have been.

CHAPTER 1
ENGLISH ROOTS, UTOPIA FOUND AND LOST

Electricity, of course, takes the place of all fires.

Edward Bellamy

The appropriation of North America by European populations was as much an energy revolution as a cultural one. The early colonists saw the New World in contrast to where they had come from, as an unspoiled place of exploitable resources where sources of energy were exceptional and unlimited. In just two centuries after their arrival, regions of the same pristine land were transformed into a high-energy, smoke-filled, urban-industrial landscape much like the Old World many Europeans had left behind.

By the time author Edward Bellamy sat at his desk in 1886 to begin writing the utopian novel *Looking Backward: 2000–1887*, a rising middle class of Americans was facing a cultural conundrum. As a society that had begun to define itself as socially and technologically advanced it remained dependent upon the fires of Old World energy. Thinking in terms of improvements, advancement, and profits, capitalists and social reformers had built a society based on production, with unlimited contrivances and unlimited energy, while at the same time maintaining a colonial worldview of pristine pastoralism. Within a growing culture of consumption, Americans had constructed an imagined environment in concert with the idea that both the land and the culture of the United States were exceptional.

As they were first identified by Tocqueville upon his description of the country in 1831, the exceptional qualities of the

United States included liberty, republicanism, and egalitarianism enclosed within a resource-rich geography, which taken together were globally unique and exempt from historical forces. While serving as a basis to an assumed economic, political, and social superiority, these ideas failed to capture what was beginning to happen in the Gilded Age air. When cities such as Pittsburgh and Chicago became as smoke-filled as Manchester and Sheffield, critical components of exceptionalism began to waver. Although the first Europeans found the New World to be a land where energy resources were plentiful and the air was smoke-free, industrialization had created an environment that mirrored the Old World. By the latter half of the nineteenth century, smoke began to obfuscate the imagined pastoral world of America. Amid a growing faith in technology that began to take hold in the latter half of the nineteenth century, an increasingly urbanized society sought solutions to the dirty energy sources that powered industry and provided the basic need for light. When simultaneous advances in electricity transformed energy from a visible to an invisible commodity, long-held ideologies of environmental and energy exceptionalism that had dimmed behind the dark clouds of burning coal began to rekindle.

As reflected in Bellamy's novel, the smoke of an energy-intensive society was a dystopian feature that did not fit into the utopian society of the American mind. Bellamy's protagonist Julian West, who regains consciousness in the year 2000 after falling under a mesmerist's trance more than a century earlier, finds himself awake in a new world rid of the controversies and issues present at the time of his sleep. As West is being familiarized with the future America, his host, Mrs. Leete, explains how society's old problems have been solved, many of them with technical solutions. In describing how electricity has taken the place of all fires, she strikes a theme familiar to many American Gilded Age utopian works that saw emerging technologies as part of a new worldview that was better, cleaner, and brighter than the past.

Bellamy's idea that electricity would replace fire was warranted at the time of his writing. Just five years earlier, in 1882,

American society's close relationship with fire was already undergoing a major change. On September 4 of that year, Thomas Edison's Pearl Street Station in Manhattan had begun to generate electricity and send it out quietly, cleanly, and invisibly through underground copper wires. This event not only was a milepost on the path to a new method for the delivery of energy, it also began to alter the way in which Americans would construct ideas about energy and their environment by dissociating fire from useful power. Through the underground mains leaving Pearl Street, clean electrons flowed indiscernibly, distancing themselves from the coal, fire, steam, and smoke that produced them. In an instant, the spatial dynamics of power generation and consumption changed, and energy became an abstraction.

When Bellamy's Mrs. Leete explains the displacement of fire by electricity, she positions fire as regressive and electricity as progressive. Considering the era's experience with energy, her conception is justified. There was nothing abstract about the human encounter with energy prior to electrification. The simple acts of providing light and heat for one's family was fraught with risks. A little more than a decade before Pearl Street's start, Catharine Beecher and Harriet Beecher Stowe wrote that careful management of coal furnaces in households was a necessity, else the devices would "poison families with carbonic acid and starve them for want of oxygen."[1] Smoke and soot, both indoors and outdoors, had been a growing problem in the mid-nineteenth century. Although some health professionals believed that smoke was beneficial, sanitarians had long held that fresh air, both indoors and outdoors, was imperative to good health.[2] In 1848, Dr. John Hoskins Griscom, one of the pioneers in public health in the United States, published *The Uses and Abuses of Air: Showing Its Influence in Sustaining Life, and Producing Disease,* which goes on at length regarding the dangers of carbonic acid and coal smoke.[3] The common thread between Dr. Griscom's and the Beecher sisters' concerns, as well as Bellamy's characterization of the environment, was fire. Before electrification nearly all energy utilized by humans, especially for domestic use for light, heat, and cooking, involved close contact with fire. While wind power and waterpower appeared in various civilizations for motive power, visible, proximal fire in some form was the energy technology around which society had organized for at least three hundred thousand years.[4]

Of course, Edison's Pearl Street Station did not eliminate fire from energy production, but it did represent part of a major paradigm shift in the history of technology. While the fire within a coal furnace represented a salient form of energy in which energy production and consumption were spatially contiguous, electricity separated the processes. With electrification, energy production and consumption became detached—wires transported usable energy and left the fire and detritus behind. Up to that time, fire had served humankind well. In the legend of *Promētheús Pyrphóros*, fire was the civilizing force for mankind, with the author Aeschylus possibly foretelling the dependence upon controlled combustion that civilization would develop. For Pliny the Elder, fire was the highest of the four elements, above air, earth, and water—identifying "Humani ignes," or fires made by men, as a wonder of nature.[5] For Charles Darwin, the discovery of fire was probably the greatest discovery ever made, excepting language.[6]

As important as fire was for human domination of the planet, its usefulness came not only from its exothermic energy in the form of heat but also from its portability. The fire pit, the candle, the cookstove, and the lantern could all be built adjacent to the need—a necessity because to take advantage of fire as an energy source one must always be in relatively close proximity to the flame. Fire is in effect a portable source of light and climate, yet it does not travel alone. Fuel must be acquired, flames must be managed, and the corollaries of flame, usually in the form of smoke, soot, and ash, are always present. These three elements of fuel, flame, and the byproducts of combustion have always been inextricably bound together, and it was this arrangement that drove the Beechers to seek solutions and Bellamy to eliminate fire completely from his utopian world. Until electrification, the energy technologies that society confronted on a regular basis had remained unchanged since Pliny the Elder mused on its place in the first century CE. Fire technologies, or pyrotechnologies, involved not just the furnace or stove but all devices and methods through which fire-based energy could be consumed.[7] The simple candle, the lantern, and the fire pit, along with devices to start, manage, and regulate fire, had gone unchanged for millennia. For all pyrotechnologies, the acquisition of fuel as well as the management of soot and smoke was a conscious, salient, and often difficult task. Collectively, pyrotechnologies were antithetical to a progressive America, with an unpleasantness reminiscent of a Dickensian world—in no way aligned with an ideology of modernity. Before Edison's Pearl Street, transportation had modernized via the railroad and communications had modernized via the telegraph, yet the simple acts of cooking, heating, and interior illumination still actively involved open flame.

Bellamy's vision was of a world without fire, period. Progressive Era futurist King Camp Gillette expressed this same sentiment through his imagined planned community of Metropolis, which would be totally electrified.[8] For many utopian authors, as well as those seeking practical solutions to issues involved with pre-electric energy, fire was an unsafe and onerous part of everyday life that needed to be either tamed or eliminated. Smoke was unpleasant, fuel acquisition was burdensome, and flame management tenuous and often unsafe. While Bellamy, Gillette, and the Beechers constructed meaning from their own experiences with fire-based energy, their negative perceptions of pyrotechnologies were not new concerns, but a continuation of ideas about energy that came before them. Just as ideas about republicanism and progressivism had crossed the Atlantic, so did ideas about technology and energy. Smoke and fire represented old-world oppression, incongruent with visions of American pastoralism and exceptionalism. Perceptions of America as "technologically sublime," as historian Leo Marx puts it, included no place for the fire, smoke, and dissonance associated with fire-breathing, industrialized England.[9]

Visions of a utopian America, where energy was clean, plentiful, and inconsequential—in an environment that was free of smoke and soot—did not originate in nineteenth-century

America; they arrived on the first ships of English colonists two hundred years earlier. Describing New England in terms of "Earth, Water, Aire and Fire," the Reverend Francis Higginson in 1630 found the air to be superior and clean-burning wood for fuel to be abundant.[10] Higginson's remarks not only illustrate the importance of plentiful, consumable energy but also indicate the importance of clean air to those from England in the seventeenth century. In his writings, Higginson places fire, fuel, and clean air prominently, and directly contrasts these elements with the situation in old England. His comments represent the earliest roots of American energy exceptionalism, an ideology characterized by an imagined environment of clear air and limitless fuel. Higginson's observation that "the temper of the aire of New England is one special thing" demonstrates his utopian valuation of a nonpolluted environment, while the phrase "we have plenty of fire to warm us" illustrates that the element of fuel for fire was not something that had been taken for granted in the past. While Higginson indeed referred to fire, he celebrated abundant wood fuel and an atmosphere void of oppressive smoke. In escaping the persecution of Puritans by King Charles I, Higginson evaded more than just the religious oppression of the Church of England; he also left behind the oppressive environment associated with the pyrotechnologies of his homeland. Long before the colonists arrived in New England, high-energy costs, shortages of fuel, and atmospheric pollution caused hardship in a dystopian England.

In English history, as early as the 1086 Domesday Book, the "Fumage," which was a tax imposed on a per-chimney basis, called for smoke-farthings in payment to officials known as chimneymen—the forebears of modern meter readers. Taxes, the rising costs of acquiring fuel due to deforestation, the loss of rights to estovers associated with enclosure, and the loss of chips for workers were just a few examples of the burdens related to domestic fire-based energy in England.[11] In addition to fuel costs, the acquisition and handling of fuel for fires were burdensome as well. Not only did the gathering of fuel and tending of fires add to the salience of energy production but fuel shortages meant that the basic need for household energy could not be taken for granted. Numerous examples demonstrate the sensitive nature of the fuel supply chain in England, as "sufficient housbote" (wood for the hearth) was often a contested resource among tenants of abbeys.[12] Agrarian records have shown that localized shortages of wood in England and Wales due to enclosure and exploitation of forests in the sixteenth and seventeenth centuries contributed directly to the tenuous nature of home energy production, forcing families to burn dried dung and furze.[13] Difficulties in the acquisition of fuel often led to rebellion as the raw materials needed for the provision of household heat for survival and cooking became a contested resource.

As burdensome as the acquisition of fuel stocks was in England before North American colonization, another salient aspect of energy production was the smoke. As early as 1285, London officials formed the first commission to investigate the airborne smoke problem as the switch from wood to fossil fuels took its toll on air quality.[14] When deforestation in the early thirteenth century drove Londoners to switch from wood to sea-coal as their primary energy source, the airborne effluents of sulfur dioxide, carbon dioxide, nitric oxide, soot, and particulate matter led to some of the earliest known pieces of environmental legislation. In 1306, King Edward set up a ban on sea-coal for some uses, with a punishment of "grievous ransom" in the form of fines. The ban was lifted in 1329 due to the demands of manufacturers and the use of coal in London increased dramatically between the fourteenth and eighteenth centuries.[15]

By the time Francis Higginson and his fellow Puritans were born, smoke and fuel shortages were already part of the intellectual discourse. In 1606, when William Shakespeare wrote *Macbeth*, his carefully crafted trochaic tetrameter "Fair is foul, and foul is fair / Hover through the fog and filthy air" directly connected evil—through the chant of the three witches—to air quality. This demonization of foul air carried forward in seventeenth-century English literature. The theme of coal smoke reappears in Sir Thomas Browne's 1674 work *Hydriotaphia*. Browne described great mists that were "hindering the sea-coal smoke from ascending and passing away" and were "drawn in by the breath... which may produce bad effects, inquinate the blood, and produce catarrhs and coughs."[16] Browne's works are demonstrative of the toxic miasma present in mid-seventeenth-century London.

Hydriotaphia provides a direct link to later American ideas about clean air and the environment. Ralph Waldo Emerson's journals reveal that he read Browne around the time that he wrote *Nature* in 1836. While Emerson's interest in Browne is in his broad analysis of antiquities and modernity, Emerson's celebration of "blithe air" also connects to Browne's musing on the "bad effects" of sea-coal smoke.[17] By 1661, when John Evelyn began to write *Fumifugium: Or, The Inconvenience of the Aer and Smoake of London Dissipated*, the sensory effects of energy generation in London were well established. Nearly three centuries after King Edward's attempt to curb coal smoke in London, Evelyn's work appeared as one of the earliest treatises to take up the subject of the air pollution resulting from contiguous energy technology. *Fumifugium* did not take issue with fire itself, but with fuel, as "the indulgence of sea-coale in the city of London" was the problem that Evelyn identified.[18] Realizing that "there is no Smoake without fire, and there is hardly any fire without Smoake," one of his proposals was to supply London with wood, which was easier said than done.[19] Wood was a rare commodity in London by the 1660s, a consequence of overcutting and overuse. Timber was available in other parts of the country, but deforestation near London by the time of Evelyn's writing led to fuel shortages that made it prohibitively expensive for Londoners to turn wood into an energy source. Already by the time of Evelyn's writing, Sir John Wintour was busy perfecting a method for the calcination of coal, for which he petitioned King Charles II for a patent. Wintour's patent, which was to char and calcinate coal "as to make it an excellent fuel without smoke" is indicative of early attempts to separate smoke from combustion.[20] Evidence suggests that despite Wintour's invention and Evelyn's proposals, Londoners often viewed their environment in terms of smoke.

Knowing the environmental conditions in parts of England in the seventeenth century allows one to better understand Higginson's musing that "The temper of the aire of New England is one special thing." His comments regarding the cost and difficulties in the acquisition of energy are also telling, as Higginson points out that in New England a "poor servant is able to afford . . . more good wood for Timber and Fire than Noblemen in London."[21] Additionally, Higginson's remarks on the ease and abundance of energy in the New World were not limited to wood, as he recognized other potential fuel sources for warmth and illumination. The abundance of fish allowed for the affordance of "Oyle for Lamps," and pitch-soaked "Pine Trees burn as cleere as a torch."[22] While one cannot generalize from Higginson's writings alone, ideas about clean air and ample fuel were commonly found in the writings of early colonists. In John Smith's *Generall Historie of Virginia, New-England, and the Summer Isles* of 1616, "the wholesome aire" and "plenty of . . . Wood" were features in this "pleasing country."[23] In celebrating "huge bon(e) fires of sweet wood," Smith clearly appreciated the abundance of fuel and clean air found in the New World—and viewed the resources and environment as his for the taking.[24] The rhetoric of an unspoiled environment with ample fuel recurs in Samuel Purchas's remarks in *Purchas His Pilgrimes*, of 1625. By recognizing the "healthfull" air in America where both inhabitants and cattle are "fruitful and grow in a good or better manner" from whence they came, Purchas expresses ideas similar to those of Smith and Higginson.[25] This sentiment was echoed by Louis Hennepin, Louis Joliet, and Jacques Marquette in 1699 as they explored the upper Midwest, who noted that the "air is temperate and open," and "forests of all manner, especially oak[, which could be] brought over to Europe, whereas [the forests there] are exhausted."[26]

Within the narratives of the Europeans, there was both an appreciation for the clear air of North America and a desire for the abundant fuel, which were prerequisites for colonization and exploitation. At the core of the European energy philosophy was the potential for selling excess timber for profit and utilizing the great stocks to build and manufacture. Enclosure of North American lands meant enclosure of North American energy, as trees represented the first great North American fuel source. Under the European model resources were "merchantable commodities," there for the taking—beyond subsistence—as a source of potential profit.[27] The enclosure and exploitation of energy resources, simultaneous to the celebration of the pure environment, was the beginning of a distinctive component of American energy exceptionalism. Despite the providential contentions of the first Europeans, however, the lands that they encountered were not free of smoke, nor were the ample sources of fuel untapped. Indigenous cultures utilized wood fires for warmth, light, hunting, and production of metals, yet they were not excessive energy users with a penchant for surplus materials and profit.[28] As characterized by William Cronon, the indigenous peoples of New England needed diversity and mobility, which led to an avoidance of acquiring any "surplus property."[29] In the case of Native Americans, energy was used as needed for subsistence; for the Europeans, surplus fuel resources were a source of revenue.

The descriptions of easily accessible fuel and a smoke-free climate in the writings of Higginson, Smith, Purchas, and the French explorers indicate that they considered the New World to be the antipode of the Europe they had left. While the motives of the early colonists are classically understood within traditional historical narratives as a search for profits or an escape from religious persecution, the writings of these settlers indicate that a search for abundant energy and clean air are viable categories for analysis. From this perspective, the historiography of energy and the environment is a narrative of resource and environmental declension. As pyrotechnologies depleted the prime fuel supplies and fouled the air in England, colonists sought new sources of energy and clean air in the New World. The writings of Higginson, Smith, and Purchas speak of an exceptional land—with descriptions of "wholesome aire" and "good wood for timber." Three centuries later, Bellamy and other utopian authors similarly described their imagined worlds. In his 1894 utopian novel, *A Traveler from Altruria: Romance*, writer William Dean Howells describes a new land where "the air is so glorious I don't mind losing a night's sleep."[30] Like Bellamy, Howells imagined an environment that he longed for—a paradise that had been lost by the 1880s—not unlike the world that the colonists found in New England in the 1600s. In a span of three centuries, from the English writings of Browne and Evelyn to the journals of Higginson and Smith to Bellamy and Howells, the authors describe the negative consequences of energy production. While the first European colonists to North America may have escaped the world of fuel shortages and smoke that they associated with their homeland, it would not take long for visions of England to reappear.

CHAPTER 2

THE ENERGY REVOLUTION AND THE ASCENDANCY OF COAL

The production and consumption of fuel in any country may be taken as a full measure of its manufacturing prosperity and even of its civilization; for it is well known that as mankind advances in knowledge, their increased wants demand the increased use of heat. An all-wise Creator, with his infallible wisdom, foresaw this, and placed under the earth's crust, and yet within the reach of man, a bountiful supply of the mineral known by the generic name of Coal.

Israel W. Morris

Foul air and black smoke, reminiscent of the England that Francis Higginson and other colonists left behind, were never part of the American imagined environment. Although the United States began as an agrarian nation, by the early 1800s an economic transition was well under way. While historians have characterized this period of US history in terms of a market revolution, it was also the beginning of an energy revolution as the quantity, sources, and characteristics of consumable energy changed.[1] Driven by industrialization and an expanding economy, the nation began its journey to becoming the number one energy consumer on earth while switching from somatic and natural energy sources to fossil fuels in the form of coal. As capitalists developed and promoted a comprehensive infrastructure from mining to transport, coal-based energy gained a technological momentum and eventually became established as the nation's primary energy source.[2] While coal was provisionally accepted by society, it created dissonance as Americans had to reconcile their desires for advancement and growth with ideas about pastoralism and energy exceptionalism.

Prior to the seventeenth century, organically powered nomadic and agricultural societies left the air in North America comparatively clean and clear when the first English colonists arrived. The energy needs of the colonies, for basic trades and domestic survival, utilized somatic energy along with biomass—in the form of wood, which grew gracefully atop the world's largest stocks of undisturbed coal.[3] The relatively smoke-free agrarian environment that the colonists found became embedded as an integral part of utopian American imagery. To the early colonists, the unadulterated pure air of the New World was the antithesis of the industrialized high-energy England that they had left behind. From the first colonizers' visions to Thomas Jefferson's ideal of the United States as a pastoral place of yeoman farmers, the imagined future was one of an unspoiled, factory-free environment.

Jefferson's idealistic agrarian doctrine was not without precedent. Between the Europeans' conceptualization of the New World and surveyor Lewis Evans's prediction that by 1775 wealth and power would "naturally arise from the culture so great an extent of good land, in a happy climate," ideas about America the garden, complete with a clear climate, were well established.[4] The pristine environment that the first colonists encountered remained part of an American imagined environment, which not only appeared in the form of Jeffersonian pastoralism but would reappear in the Progressive Era's conservation movement. When Jefferson proclaimed in 1785, "For the general operations of manufacture, let our workshops remain in Europe," he was distancing the young nation from the fire and smoke of the factory and proposing for America a low-energy, exclusively agrarian society.[5] Jefferson was expressing ideas about space and place—separating production from consumption, akin to the modern idea of "not in my backyard."

Jefferson's views about industrialization, shared by John Adams, became the subject of great debate between himself, Adams, Alexander Hamilton, and the grand early promoter of American industry Tench Coxe. Coxe took a more nuanced view of Jefferson's agricultural pastoralism, opining that manufacturing worked hand in hand with agriculture, as he expressed in his *An Inquiry into the Principles for a Commercial System for the United States* from 1787.[6] Where Jefferson's idealistic view of an American utopia was one free of industrialization and powered by muscle and wood, Coxe believed that industrialism and agrarianism not only could coexist but would complement each other. While the nation's industrial capitalist development would have proceeded without the input of Jefferson or Coxe, their debates reflect how Americans eventually constructed ideas about technology, energy, and the environment—that the United States could have it all: a pristine pastoral environment alongside a high-energy, industrialized, consumer-based society. With the passing of its first major act under the newly ratified constitution, the Tariff of 1789, the United States Congress in essence passed its first—albeit indirect—energy bill. As the act specified that part of the tariff's purpose was for "the encouragement and protection of manufactures," it became the harbinger of industrialization and the beginning of what would become a high-energy regime.[7] By the beginning of the nineteenth century an American industrial revolution was well under way. Glass and iron making had been established in Pittsburgh in the late 1700s, and Samuel Slater's first textile mill began manufacturing in Rhode Island in 1803 along with a host of other regional manufactories. Slowly, the somatic energy of an agricultural society began to be supplemented by the industrial energy sources of fire and flowing water.

Despite his visions of a pastoral America, Jefferson accelerated manufacturing with the Embargo Act of 1807. As British imports to the United States were cut off prior to the War of 1812, domestic manufacturing increased to meet the demands of a society poised to consume. By 1809 mechanization and manufacturing became unstoppable forces. Writing to industrialist P. S. DuPont de Nemours, Jefferson expressed his realization that "the spirit of manufacture has taken deep root among us, and its foundations are laid in too great expense to be abandoned."[8] While the imagery of the pastoral remained in the American mind, the roots of a dissonant high-energy society were firmly planted. When Jefferson remarked in 1814, "Our enemy has indeed the consolation of Satan on removing our first parents from Paradise: from a peaceable and agricultural nation he makes us a military and manufacturing one," he marked the transition between a society built upon agrarian energy to one built upon high consumption of industrial energy.[9] The Arcadian paradise—the pastoral utopia conceptualized first by the Puritans—was on its way to becoming a world reminiscent of Evelyn's *Fumifugium*.

The energy that powered the early periods of antebellum America came from a variety of sources. Households burned wood and a variety of lamp oils for illumination, and water along with wood powered industry up until the mid-nineteenth century. Industrial processes that required flame or forge, such as the glass and iron industries, utilized wood almost exclusively up until at least the same time period, while mechanized industries such as textile manufacturing ran on waterpower until midcentury. Although domestic coal began to enter markets as early as the 1830s for some applications, the nation's transition to fossil fuels would be gradual. Because of easy access and the vast reserves of forest, as the nation transitioned from a predominantly rural agricultural economy to one of market-based industrialization, it remained fueled by wood. Although the production of metals rose substantially during the same period and English iron production was reliant on coal, the United States continued to be reliant on timber. The domestic iron and steel industry's continued use of wood-derived charcoal for steel production—as opposed to coal-derived coke—lasted until late in the nineteenth century.[10]

Before 1850 the steam engine, whether fueled by wood or coal, had not yet taken command, and US manufacturing was powered almost exclusively by water. Statistics derived from the 1832 McLane Report, which was compiled by then–Secretary of the Treasury Louis McLane, indicate that only four large industries out of a total of 249 capitalized at $50,000 or more utilized steam power.[11] This excludes Pittsburgh's firms, which were, in the words of business historian Alfred D. Chandler Jr., a "striking exception" due to the types of industry and their proximity to large deposits of bituminous coal.[12]

Waterpower's dominance in America's early industrial revolution came about because it was inexpensive and provided easily exploitable mechanical power. While England's manufacturers had adopted usable steam power before the United States, the abundance of water energy along with an undeveloped coal mining and transport infrastructure delayed the nation's transition to coal and steam power. Prior to the 1850s it was not yet economically feasible to ship large amounts of coal to the nation's manufacturing centers, which were at that time concentrated in the Northeast.[13] The water-powered manufacturing enterprises in New England were larger and newer than any textile operations that had come before, in England or in the United States. As such, descriptions of mills such as those at Lowell, Massachusetts, reveal how society constructed ideas about energy in general and smoke in particular. While descriptions of an already smoky Pittsburgh included monikers such as "the Birmingham of America," hydropower-driven cities such as Lowell were viewed in starkly different terms.[14] The memoirs of Alexander Mackay from the 1840s provide a depiction of Lowell that is diametrically different from those of Pittsburgh. Mackay writes, "On approaching Lowell, I looked in vain for the usual indications of a manufacturing town with the tall chimneys and the thick volumes of black smoke belched forth by them . . . and I was struck with the cleanly, airy, and comfortable aspect of the town."[15]

Where Pittsburgh was defined by its smokiness, Lowell was seen as a commercial paradise, void of fires and the blackening effects of coal. If Lowell represented a utopian future—a manufacturing center that ran without smoke—Pittsburgh represented the past, a throwback to Manchester or Birmingham of the Old World. For industrial energy, the waters of the Merrimack represented what historian David Nye describes as the "moral machine," a new and better way than what was found in England.[16] Colonel Davy Crockett's account of his visit to Lowell in 1835 parallels Mackay's comments and provides credence to Nye's interpretation. Not only did Crockett find the environment Eden-like, from the moment he began his trek from the headwaters of the Merrimack eastward into town he made nothing but positive comments about his experience. The cleanliness of the town, the "happiness and prosperity," were all striking to Crockett.[17]

In contrasting the characterizations of Pittsburgh and Lowell, an early picture emerges of how energy technologies were conceptualized. The visible energy that drove Pittsburgh, manifested in smoke, created an association with England and old-world technology, while the smoke-free Lowell was seen as American improvement. Although industry dominated both cities, Pittsburgh was the environmental "other," described by Charles Dickens as being "like Birmingham in England" upon his touring the city in 1842.[18]

It is here that socially constructed images about US energy technology begin to mirror the ethos of an imagined American exceptionalism as it applied to the land and the environment. Englishman Anthony Trollope found the water-powered factories in Lowell a "realization of a commercial utopia."[19] Social commentator Harriet Martineau found none of the natural scenery "deformed by the erection of mills" in Lowell when contrasting the American manufacturing center to that which existed in her industrialized homeland of England.[20] Martineau's description of Lowell reinforces how American exceptionalism applied to the young nation's industrialization: "As to the old objection to American manufacturers, that America was designed to be an agricultural country, it seems to me . . . that America was meant to be everything."[21] Dickens commented of Lowell that "in all, there was as much fresh air, cleanliness, and comfort, as the nature of the occupation would possibly admit of."[22] While Martineau, Trollope, and Dickens were passing commentary on the system of manufacturing at Lowell that they found superior to that in England, their observations, along with those of others, explicitly note the lack of smoke.

The fresh air of Lowell, along with the technical savvy of water-powered cotton mills, reflected an ideal state of technology with its roots in American millennialist thought. While the Jeffersonian vision of the country may have rejected old-world, British-style industrialization, advancements in American manufacturing technology that were seen as "improvements" fit well into the ideal model of the New World. British manufacturing, with its smoke, steam, and oppression, was regressive, and American manufacturing in Lowell was clean, orderly, and progressive—an exceptional break from English cities like Manchester and Sheffield. Late in the eighteenth century, millennialists, including Joseph Bellamy and Samuel Hopkins, followed the views of Jonathan Edwards, who spoke of improvements and contrivances as part of the New World utopia. On manufacturing, Hopkins wrote: "There will also doubtless be great improvements and advances made in all those mechanic arts, by which the earth will be subdued and cultivated, and all the necessary and convenient articles of life, such as all utensils, clothing, buildings, will be formed and made in a better manner."[23] Edwards, Bellamy, and Hopkins carried not only significant religious influence in the early nineteenth century but also secular influence where postmillennial beliefs complemented ideas about national progress and technological advancement.[24]

While the sophisticated mechanical elements of Lowell fit with the foundations of technological sublimity, the lack of smoke was also in harmony with the positive hopes of a new America. Although air pollution and environmental issues were not yet part of the public discourse, fresh air was valued as a vital part of health and well-being. Clean, smoke-free air was backed by health reformers such as the transcendentalist Amos Bronson Alcott and the Christian physiologist Sylvester Graham, who championed the early clean living movement.[25] While waterpower was by no means a new technology, in Lowell it took on a new identity as an alternative to smoke and fire. Even at the nation's early stages of industrialism, visible, smoky pyrotechnologies could be construed as the antithesis of progress. Whether evidenced by the positivism associated with Lowell's clean air, the sheer number of smoke- and spark-controlling devices, or the number of inventions for harnessing wind power and waterpower submitted to the US Patent

Office in the first half of the nineteenth century, the impulse for cleaner energy was already present.[26] One of the more grandiose efforts came from social utopianist John Adolphus Etzler. Etzler came to the United States in 1831 as a member of the Muhlhausen Emigration Society under the sponsorship of John Augustus Roebling, who would go on to build the Brooklyn Bridge beginning in 1870. Etzler's relevance to the history of energy in the United States comes from his notable stance on clean, renewable power. In his 1833 work *The Paradise Within the Reach of All Men, Without Labor, by Powers of Nature and Machinery: An Address to All Intelligent Men*, Etzler wrote: "The basis of my proposal is, that there are powers in nature at the disposal of man, [a] million times greater than all men on earth could effect, with their united exertions, by their nerves and sinews. If I can show that such a superabundance of power is at our disposal, what should be the objections against applying them to our benefit in the best manner we can think of?—If we have the requisite power for mechanical purposes, it is then but a matter of human contrivance, to invent adapted tools or machines for application."[27]

Etzler theorized that the wind, sun, and tides could provide ample power for society, and attempted to construct a number of wind power devices, none of which came to commercial fruition.[28] Regardless of Etzler's lack of success, his importance here is that along with those who saw the lack of coal smoke at Lowell in utopian terms, such as Trollope, Dickens, and Martineau, Etzler too saw coal as anti-utopian and under the control of "vicious energy monopolies."[29] In his brief and inconspicuous appearance in American history, Etzler not only saw the possibilities for sustainable energy but also foretold how corporate interests would eventually come to control the country's most critical commodities.

Others after Etzler continued the quest for renewable, clean energy systems, the most notable being John Ericsson, the American who developed and built the famous ironclad the USS *Monitor*. Ericsson argued that coal was not a sustainable source of energy, stating that "the time will come when Europe must stop its mills for want of coal" and theorizing that "Upper

Egypt, with its never-ceasing sun power, will invite the European manufacturer" to utilize solar power.[30] Yet by the time Ericsson developed his heat engine in the 1860s, coal had already supplanted waterpower as the primary energy source powering America's industries. As utopian as smoke-free energy may have been to social critics at the time, manufacturing was driven by profits, and waterpower was only in place because of its abundance and low cost. If waterpower, along with Etzler's and Ericsson's early work on wind and solar power, represented an early impulse for the development of a society based on sustainable energy, any possibilities for future development of clean power passed in the mid-1800s, as fire-based energy fueled by coal became less costly—and thus a more profitable energy source for manufacturers.

The rise of coal and the subsequent beginnings of an American society based on fossil fuels are not attributable to a single cause. From a cultural standpoint, Americans' demands for more goods and services created an industrial economy that would challenge England's by the middle of the nineteenth century. As demand for goods increased, conversion to steam power, ultimately fueled by coal, became necessary as the scale and location of manufacturers outstripped the energy that waterpower could supply.[31] Although early steam power in the United States was fueled by wood, as coal mining, distribution networks, and steam engine technologies developed, coal became the fuel of choice. Fundamentally, coal is a much more energy-dense fuel than wood, and the inherent efficiencies gained—more heat per unit of weight—made it a superior fuel over biomass. While the English had been using coal for centuries, the geography of the United States made transport initially difficult, and until reliable networks were established, wood and water that were near the point of consumption were the most cost-effective means of power. Ralph Waldo Emerson identified a key attribute in the evolution of the nation's energy usage: "Coal lay in ledges under the ground since the Flood, until a laborer with pick and windlass brings it to the surface. We may well call it black diamonds. Every basket is power and civilization. For coal is a portable climate. It carries the heat of the tropics to Labrador and the polar circle."[32] With energy portability came flexibility of use—energy for manufacturing and domestic use could be delivered to where it was needed. Wood was not cost-effective to transport long distances, and with a superior energy-to-weight ratio, coal's portability was a critical factor in its adoption.

As with coal itself, the origins of the US coal industry are found in many corners. Whether it is the pure and clean-burning anthracite coal concentrated in eastern Pennsylvania, or its much dirtier, smokier cognate, bituminous coal, concentrated in western Pennsylvania, West Virginia, and Ohio, a coal supply network began to emerge in the first half of the nineteenth century. While Pittsburgh had been fueling its forges and furnaces with coal since the mid-1700s, markets farther away from the mines did not begin to adopt the mineral fuel until after the first quarter of the nineteenth century. In Ohio, mine owners aggressively promoted coal as an alternative to wood for both industrial and domestic use by the early 1830s. In Meigs County, deed transfer archives show an 1804 transfer from "E. Gerry & wife" to S. W. Pomeroy, a businessman from Boston, which took place in what was then Salisbury Township, Ohio.[33] Pomeroy went on to form the Pomeroy Coal Company, which was one of the earliest in the state. Coal had been mined in the township prior to Pomeroy, but he and his sons promoted and shipped coal down the Ohio River on a large scale, sending over 116,000 bushels to Cincinnati alone by 1818.[34]

Prior to that time, domestic users preferred wood fuel, but aggressive marketing, which guaranteed price and promoted coal as a superior fuel to wood for household use, proved to be a successful campaign for the company.[35] To the north, Cleveland's market for coal was growing around the same time, with most of its coal coming from mines in Tuscarawas County. At least eighty-seven thousand bushels of coal were being shipped annually to Cleveland from mines in New Castle, Ohio, in the 1830s.[36] Both Cincinnati and Cleveland became major destinations for southern Ohio coal as industrial and domestic use expanded into the mid-1800s. As bituminous mining and distribution took shape in the West, the market for anthracite coal in the East was emerging as well. The earliest substantive phase in the rise of anthracite coal is well documented in several historical accounts, one of the most comprehensive being found in the work of historian Alfred D. Chandler Jr., who hypothesized that mine owners and entrepreneurs "perceived a market for anthracite, developed the technology for its use, and built canals for its transportation."[37] Chandler's analysis identifies the iron and steel industry east of the Alleghenies as the first to adopt anthracite in manufacturing in the 1830s. Chandler cites the emergence of canals and eventually the new railroad as the beginning of a new energy network; this in turn drove technological momentum for the use of coal as new processes and devices began to cater to the transition away from wood.[38]

While the iron industry consumed much of the coal both in Ohio and in the East, coal-derived energy accelerated more rapidly with the gradual adoption of steam power. The significance and history of the steam engine has been well documented. For Karl Marx, the steam engine represented the first time that motive power "begot its own force by the consumption of coal and water . . . [and] was entirely under a man's control."[39] Others spoke of the steam engine in pure ethereal terms. For Robert Henry Thurston, distinguished professor of mechanical engineering at Cornell, it was related to the divine: "As religion has always been, and still is, the great moral agent in civilizing the world, and as science is the great intellectual promoter of civilization, so the Steam Engine is, in modern times, the most physical agent in that great work."[40]

Although the steam engine's significance in modernity is well known, its significance in how energy is conceptualized has little precedence. By converting the energy of fire to mechanical energy, which was distanced from the fire through shafts and pulleys, the steam engine places a layer of abstraction between fire and useable force. Not only did the steam engine begin to modify the perceived spatial dynamics of energy; it accelerated a locational shift as well. When utilizing waterpower, industry has to move to the energy source, which in many cases at the time was rural and away from markets.[41]

With coal power and the steam engine, the energy source could be moved virtually anywhere. For manufacturers, this meant that production could move closer to markets and closer to shipping routes. For transportation, steam technology was truly revolutionary, as it allowed for self-propelled travel first via water and then by rail. Much like coal was significant in its ability to provide a portable source of heat, the development of steam power marked the beginnings of portable mechanical energy. In the history of energy, no development was as critical as portability. While the steam engine still represented a salient energy technology in which energy production and consumption remained proximal, its ability to convert heat to motion wherever needed represented the beginning of a major paradigm shift in the way energy was conceptualized and commoditized.

Although Taqī al-Dīn, Jerónimo de Ayanz y Beaumont, and Thomas Savery were the early pioneers of steam-powered automata, it was Thomas Newcomen, followed by Matthew Boulton and James Watt, who developed the first truly viable engines in England in the early 1700s.[42] While in England Boulton and Watt steam engines began to make hydropower obsolete by 1800, the United States was slow to adopt steam power.[43] In the United States, Oliver Evans invented a high-pressure steam engine by 1801 that was used on riverboats and in some flour mills, but it lacked the efficiency needed for wider adoption.[44] From all indications, it was not until the introduction of the American-made Corliss steam engine in 1849 that the economics of steam power in the United States began to change. The Corliss may not have been the first steam engine available in the United States, but it was the most cost-effective to run, and as such became a viable alternative to waterpower. Fitted with rotary valves and variable valve timing, the Corliss engine offered the best thermal efficiency and made steam power more economical and able to compete with the waterwheel on a basis of cost.[45] In one particular instance, at the James Steam Mill in Newburyport, Massachusetts, the Corliss engine consumed 5,690 pounds of coal per day, about half compared to the Bartlett engine the mill had been running, which consumed 10,283 pounds of coal per day.⁴⁶ This not only kept the cost per horsepower below previous steam technologies but made the Corliss a feasible competitor to waterpower.

Recent studies have found that steam engines in general, and the Corliss design in particular, played a major role in the shift away from water-derived energy in the second half of the nineteenth century, which in turn had a major effect on industrial demographics. As the Corliss allowed for economic power to be derived at virtually any location near fuel supplies, it served as a major catalyst for industrial relocation and urbanization, along with the development of stable railroad networks.⁴⁷ The widespread industrial adoption of coal-based energy in the last half of the nineteenth century not only marked the beginning of a US society powered on fossil fuels but also accelerated problems associated with smoke, fire, and energy. As both industry and population grew, America transitioned from rural to urban and away from wood and water as its primary sources of energy. At the same time, immigration and a natural population increase led to an overall increase in energy consumption. Demand for iron, steel, and manufactured goods grew rapidly as the nation's infrastructure was being built. As industrial users adopted coal, which affected outdoor air quality, indoor air quality also became problematic, as the types of energy used and the ventilation in domestic and commercial buildings varied considerably. While fuel for household heating and cooking gradually shifted from wood to coal, fuel for illumination varied from candles to a variety of lamp oils. Whale oil and a variety of vegetable oils were most common in the early 1800s, and whale oil use continued well into midcentury, with over six million gallons consumed in 1841.⁴⁸ Camphene, a lamp fuel made primarily from pine oil, was adopted slowly due to a high incidence of explosions, an inconvenient and quite salient side effect of illumination via proximal flame. Eventually, lamps designed to use camphene cut down on risks and the distillate was used alongside other lamp oils.⁴⁹ The inconvenience of pyrotechnic domestic energy knew no class boundaries. From Jacob Riis's accounts of smoky and dirty tenements with "oil stoves that serve at once to take the raw edge off of the cold and to cook the meals by," to parlor guides that discuss "the annoyances of soot and smoke," issues with energy derived from fire indoors were as problematic as coal smoke became outdoors.[50]

While historical accounts of various shifts in technology and population help explain transitions in both energy consumption and characteristics, the extent of these trends becomes clearer through statistics. According to the US Census Bureau's *Historical Statistics of the United States 1789–1945*, four hundred thousand units (1,000 board feet) of standing timber were cut in 1809, and that figure jumped to eight million units by 1859. While the census records for wood usage do not delineate lumber and firewood, other studies have shown that the overall consumption of wood for fuel accelerated substantially. According to a study by geologist Michael Williams, an estimated 113,740,000 acres of forests were cleared in the United States prior to 1850, and in the ten-year period between 1850 and 1859, over 39,705,000 more acres of forests were cleared. Much of the wood went for fuel.[51] Sources showing relative usage of wood and coal for domestic use in selected regions indicate that wood accounted for nearly 75 percent of domestic fuel, with coal making up the balance in the 1820s.[52]

Documented coal production in the United States was 15,000 net tons between 1807 and 1820, which rose to a total of 14.5 million tons by 1860.[53] Coal production climbed to 40 million tons by 1870, 79 million by 1880, and 157 million by 1890, the year that the US per-capita consumption of coal topped that of Great Britain.[54] Considering only the burning of coal, and that even the most complete combustion of coal yields approximately 2.8 tons of carbon dioxide for every 1 ton of coal burned, the United States was emitting over 40 million tons of carbon into the atmosphere annually by 1860, a number that would jump to 439 million tons of carbon dioxide emitted by 1890.[55] While Pittsburgh was the lone city associated with smoke early in the nineteenth century, cities such as Cincinnati, Cleveland, and Chicago began to share the honor by the mid-1800s. By 1860 there were reports of "smoke [that] pervades every house in Cincinnati, begrimes the carpets, blackens the curtains, and worries the ladies," along with associations of Cincinnati with the Old World: Cincinnati's "fuel is identical in effect . . . [to the] smoke-begetting coal that gives to the English town its grimy, inky, hue."[56] In Cleveland, the Division of Health was declaring that "smoke from coal [was] unpleasant and possibly unhealthy," while others such as real estate dealer L. E. Holden were seeking to satisfy the demand for "elegant homes away from the smoke and dust of the city."[57] Holden's comments begin to illustrate emerging ideas about energy production and space—that living away from smoke-producing city centers was becoming a desirable proposition. In Chicago, a citizens' committee on smoke began reporting periodic "smoke observations," which documented individual industrial smoke releases listing company name, type of coal use, mode of firing, origin of coal, and condition of smoke. A typical listing was "Wisconsin Elevator Company, careless firing, Pittsburgh coal, and black."[58]

In a period of about two centuries, the clean air that the colonists encountered upon landing in the New World changed dramatically. As the nation industrialized after American independence, an energy revolution began that was characterized by a shift from a society based on subsistence energy derived from somatic and natural sources to one of high energy consumption that ran on fossil fuels. While early industrial energy technologies driven mostly by hydropower were congruent with an ideological orientation that saw the young United States in terms of the pastoral, the emerging shift to coal-based energy became antithetical to ideas about American technological and environmental exceptionalism moving into the Progressive Era.

CHAPTER 3
THE CONUNDRUM OF SMOKE AND VISIBLE ENERGY

> Our coal measures are simply inexhaustible. English coal-pits, already deep, are being deepened, so that the cost of coal mining in Great Britain is constantly increasing, while we have coal enough near the surface to supply us for centuries. When storing away fuel for the ages, God knew the place and work to which he had appointed us, and gave to us twenty times as much of this concrete power as to all the peoples of Europe.
>
> Josiah Strong

In the latter half of the nineteenth century, Americans encountered smoke and soot—the airborne remnants of industrial and domestic energy usage—on a regular basis. Stuck in a quandary between economic growth, technological advancement, and American environmental exceptionalism, smoke became an inconvenient byproduct of progress that could not be ignored. The social response to smoke manifested itself in a variety of ways. Municipal governments as well as social commentators initially ignored the issue that they could not resolve, industrialists attempted to deny that a problem existed or saw smoke as a benefit to society, and social reformers and property owners actively worked to abate the nuisance. As smoke and soot became more acute, the vast majority of urban dwellers came to see them as a nuisance.[1] Airborne pollution issues cut across class and gender lines, and led to alliances between social groups who worked together to mitigate smoke and its associated issues.[2] Whether it was middle-class reformers or a downtrodden urban working class, antismoke crusades eventually came to be broad-based. Through an investigation of the historical record that examines both the social and the technological responses to smoke, it becomes clear that smoke from coal and other sources affected wide swaths of society. The working poor used coal in tenements, resulting in soot-covered walls as captured in Jacob Riis's images, and the middle class bought it, stored it, and dealt with ventilation as prescribed by Catharine Beecher and her sister Harriet Beecher Stowe. Looking to eliminate smoke were those whom historian David Stradling refers to as "urban reformers," mostly consisting of middle- and upper-class women.[3]

A review of the works of influential social activists in the era reveals deep contradictions between ideas about energy, social and economic progress, and the environment. Although environmentalism as we now know it was not within the purview of those writing in Gilded Age America, several works demonstrate how the looming nuisance of smoke could not be resolved against ideas about divine providence and progress, and was not addressed as a significant social problem. One example is found within the works of Josiah Strong. An influential leader of the Social Gospel movement in the late 1800s, he was by all measures an erudite individual who wrote prolifically about what he saw as the evils of industrialization. From the production of luxuries to a xenophobic fear of cities and their "Romanizing" element of immigrants, Strong warned that the nation's downfall was well under way.[4] Though he was alarmed by the number of saloons, poor living conditions, and declining morals, Strong never made the connection between industrial "progress," exploitation of natural resources, and the degradation of the environment. He praised the nation's inexhaustible coal supplies and the nation's climate simultaneously, and in doing so revealed the conundrum within American attitudes about resources, progress, and the environment.

To Strong and others, natural resources were a gift from God for Anglo-Saxons to use as they saw fit. As the nation's coal supplies were "inexhaustible," the nation's climate was conceptualized as resilient, with air that had a "stimulating effect" that led to even more progress.[5] Strong was not alone in his characterization of coal as divine. His contemporary, Washington Gladden, also saw the nation's vast coal supplies as "the handiwork of God" and at the same time stressed that it was an employer's responsibility to provide "wholesome air" for those who labored.[6] Gladden emphasized the importance of proper ventilation and a healthful environment while praising the invention of steam engines and a variety of other automata as synonymous with deific advancement. Both Strong and Gladden saw the use of natural resources such as coal as progress while at the same time praising automation and stressing the importance of fresh air. Their attitudes reflect a paradoxical view about the providential use of resources and an American environment blessed by God.

The contradictions within the Social Gospel movement, the divinity of coal, Anglo-Saxon technical progress, and the wholesome climate of God's chosen land could not be resolved by Gladden or Strong. Smoke, soot, and atmospheric effluents were difficult to fit into the divine setting of the United States and were absent from the narratives identifying the evils of industrialization. Within the approximately five hundred pages of Strong's and Gladden's preeminent works warning of the dangers inherent in capitalism, the words *smoke* and *soot* are curiously nowhere to be found.[7] Strong does mention "black accretions of filth" on the walls and ceilings within tenement houses, but in the context of London, not *yet* in the United States.[8] Although the Social Gospel movement was primarily concerned with the moral issues that arose from industrialization, its praise of coal and progress along with the omission of smoke provides a window into how Americans began to construct ideas about industrialization, energy, and the environment. If coal was part of "God's handiwork," then smoke had to be part of the master plan as well, a fact that could only be reconciled by overlooking the matter or denying that a problem existed. Other community leaders began to reconcile the various benefits of "progress" and the "necessities of life," but also recognized the "loss to the people" that occurred when newly established smoke ordinances were ignored.[9]

In the courts, the difficulty of recognizing that smoke was in fact a problem represents an early tendency to ignore the smoke nuisance, not only by local governments but also as a policy issue. In 1884, the Illinois Supreme Court declined to say that the "mere emission of dense smoke without proof that it was a nuisance was in fact a nuisance per se."[10] In the case of *The City of St. Paul v. Gilfillan* in 1886, the court held that "emission of dense smoke from smokestacks or chimneys is not necessarily a public nuisance."[11] Other cases held that smoke was only a nuisance if it produced tangible injury to property, real or personal.[12]

While the easiest course of action for some may have been to disregard the issue of smoke, many municipal governments hesitated to deal with the issue even after antismoke ordinances were in place. In cities ranging from New Orleans to Saint Louis to Cincinnati, smoke discharge laws were in place for industrial polluters, but were ignored and went unenforced in many cases.[13] Perhaps more difficult than reining in the more obvious industrial polluters was the issue of residential smoke from furnaces and stoves, which was often more problematic due to incomplete combustion and widespread use.[14] While the courts, municipal governments, and social theorists initially took little action against the looming issue of smoke, the historical record exposes others who put coal smoke in a beneficial context. For some, smoke meant progress, and by the late 1800s the rhetoric of pitting coal as energy against progress in the form of commerce was in full bloom. In the era's context of commerce, the still-familiar cultural tomes that tie energy with progress and employment began to appear. The idea that "soft coal smoke is a nuisance, but . . . it always means business," is a somewhat typical statement by promoters of commerce.[15]

Throughout the Midwest and into the South, city promoters regularly tied "smoke from our factory chimneys" with prosperity.[16] For Samuel Hazard, the grand advocate of Pennsylvania and the state's official archivist in the late 1800s, coal smoke was nothing but beneficial. He expressed ideas about the "pure mountain air and healthy coal smoke and its peculiar good health" as a characteristic of Pittsburgh in the summer.[17] Hazard's interest was in promoting business in his state, so he was a smoke supporter by necessity. This sentiment of energy boosterism was later exemplified by novelist Booth Tarkington in his 1915 novel *The Turmoil*. The book's antagonist, Sheridan, calls the smoke "Prosperity," as the author states in third-person prose that "smoke is like the bad breath of a giant panting for more and more riches."[18]

While some tied smoke to jobs and prosperity, other smoke supporters denied that a problem existed at all. Colonel W. P. Rend, a millionaire coal magnate from Pittsburgh whose interest was obvious, claimed, "I will still go further and state that there is convincing evidence that soot, or carbon, or smoke, emitted and poured into the atmosphere of our city works no injury, but is, if anything, an advantage and a benefit to public health. I will maintain that this so-called evil, in its comprehensive and broadest sense, is no evil at all, and that in its most important aspects it is both a private and public blessing."[19] Rend's aggressiveness on the issue is no surprise—he is also notable for challenging a competitor to a duel over mine property rights in Pittsburgh.[20] A similar story played out in Saint Louis with E. Goddard of the E. Goddard and Sons Mill Company. Goddard claimed in 1888 that he had built three steel mills in the city and had never heard one complaint from his neighbors. He then asked hypothetically, "Do you know why St. Louis is the third healthiest city in the country? Why is that? I'll tell you why, it is the smoke. If it were not for the smoke St. Louis would be suffering from all the plagues and torments that afflict other cities."[21] While Rend, Hazard, and Goddard attempted to depict coal smoke as a benefit to communities and a symbol of progress, others believed that carbon emissions were curative and could even be valuable to health. There was speculation from several physicians that "fine carbon particles might be a beneficial aerial disinfectant."[22] Others maintained that coal smoke and soot had curative properties, especially against tuberculosis, even providing a "true immunity" against the disease by rendering tubercle bacilli inert.[23] Much of this speculation likely derived from common reports that coal miners who were exposed to coal dust (as opposed to smoke) were immune to tuberculosis.[24] This line of conjecture linking positive health effects to smoke was also found in positions that linked airborne arsenic, which was found in the smoke of smelters, to beautiful complexions.[25]

As those with special interests who used anecdotal evidence to support their prosmoke agendas worked hard to put a positive face on foul air, the visibility of smoke, along with the tangible effects of breathability and soot, could not be willed away with propaganda or speculation. The overwhelming belief and obvious reality—dating back to the 1300s in England—was that smoke was at the very least a "nuisance," and the health effects, at least in the form of coughing and irritation, were well known. Not only was smoke an inconvenient byproduct of progress, it was untidy, antipastoral, and increasingly seen as preventable during a time in the United States when efficiency and engineering symbolized advancement and modernity. As the smoke problem became more acute in American cities, the outcry for reform became more prevalent as well.[26] While the momentum of social activism against coal smoke began to build in the second half of the nineteenth century, the conceptualization of smoke as a mere nuisance was being supplemented with medical science that contended that smoke was a real health problem.

As early as 1848, Dr. John Hoskins Griscom, one of the pioneers in public health in the United States, published *The Uses and Abuses of Air: Showing Its Influence in Sustaining Life, and Producing Disease; with Remarks on the Ventilation of Houses*, which documented the dangers of carbonic acid, a byproduct of coal gas illumination, and coal smoke in general.[27] In a story from an issue of *Farmer's Magazine* from 1859, the dangers of coal smoke are expressed even more succinctly, as "antagonistic to the lung and brain, of the purity of one and the health of another."[28] Here, the author associates coal smoke not just with degraded physical health but with mental health as well. Obscuring the outside air and soiling the ground represented the visible effects—black soot on white snow was a visual indicator of the nuisance—yet the stealthier and insidious effects were within the body and even the mind.

As early as 1845, Scottish doctor James Copeland, MD, had identified "common coal smoke" as a frequent cause of asthma in the first volume of his *Dictionary of Practical Medicine*.[29] In the third volume, Copeland documented "coal gas, oil gas, carbonic acid and bi-carbureted hydrogen," all byproducts of combustion, as poisons.[30] Although Copeland practiced in Britain, he influenced American medicine and served as an honorary member of the Massachusetts Medical Society.[31] By 1849, medical journals such as the *American Journal of Medical Sciences* had identified smoke from coal and wood as "at all times, injurious to health . . . unless carried away and diluted in the upper regions of the atmosphere."[32] Although the medical community had identified health issues associated with smoke some forty years prior to court decisions that were questioning whether or not smoke was a nuisance, they also believed that smoke was rendered harmless once it had reached the upper atmosphere. This idea of atmospheric cleansing—or at least atmospheric tolerance—is a typical manifestation of the ideology of environmental exceptionalism that remained deeply rooted in American culture.

While smoke pollution most visibly affected outdoor air quality, indoor air quality was a concern as well. Smoke from poorly ventilated indoor stoves and fumes from coal-gas lanterns were well-publicized issues. In New York tenements, where lanterns and candles were the only source of illumination, and coal was the main source of heat, rooms "blackened with smoke" and with little ventilation were not uncommon.[33] And suffocation from the byproducts of combustion did not occur only in tenements. A *Chicago Tribune* report from February 17, 1874, with the headline "The Lake Tragedy: The Children Probably Suffocated by Coal-Gas" was typical of news in a number of cities.[34] Deaths associated with indoor heating and illumination were common and cut across class lines.[35]

Beyond living spaces, coal smoke and fumes from coal gas were problematic for storeowners. The soot, odor, and heat associated with lighting retail establishments had been a major concern and required regular cleaning and airing out of goods.[36] In Chicago, department store owner Marshall Field estimated that his "soot tax"—that is, the cost of cleaning merchandise—exceeded the cost of his real estate taxes.[37] In New York, the "nuisance in shops" due to the burning of coal gas for illumination had been a subject of discussion prior to 1870.[38] Retail establishments in Cleveland reported losses of well over 10 percent of their profits on white goods due to "the presence of coal smoke and gas" from outside air and inside lighting. Once smoke was absorbed by fabrics, storeowners found items "beyond redemption." One retail establishment in Cleveland reported a cost of $1,800 for cleaning and redecorating as well as a $2,000 bill for window cleaning and $1,500 for laundry, all due to smoke.[39]

Awareness of air quality and coal smoke was not limited to health care professionals and storeowners in the 1800s. As with earlier reform movements in the United States, women played a vital role in recognizing the issues associated with coal energy and coal gas illumination. Environmental historian Angela Gugliotta finds that although the poorest women were most heavily exposed to environmental pollution in the late 1800s, "elite" women were affected as well, and that in Pittsburgh, class barriers broke down in antismoke activism.[40] In 1869, the Beecher sisters warned that coal furnaces could "poison families with carbonic acid and starve them for want of oxygen," which is indicative of the increasing concern over indoor air quality during the period.[41] Beecher and Stowe's *The American Woman's Home: Or, Principles of Domestic Science* could be considered a primer on indoor air quality itself, with four entire chapters devoted to proper ventilation, coal burning, chimney drafts, illumination, and other concerns relating to safe energy use in the home. Other women's publications echoed their advice: *Godey's Magazine* emphasized the importance of a "proper draft" in rooms, and *Everyday Housekeeping* advised that improper drafts led to smoke and smoke is "fuel wasted."[42]

In Cleveland, Chicago, Cincinnati, Saint Louis, Pittsburgh, and other cities, groups of reformers drew public attention to air pollution problems both indoors and outdoors.[43] The Women's Health Protective Association of Allegheny County took up the fight against smoke in Pittsburgh, as did the Chicago Citizens' Association. In Saint Louis, a social club of women known as the Wednesday Club created and supported an antismoke group known as the Citizens' Smoke Abatement Association, which lobbied for antismoke ordinances in 1893. In Cleveland, there were several antismoke groups, including a fine arts group and the Society for the Promotion of Atmospheric Purity.[44]

The issues Cleveland faced were typical of other manufacturing cities in the late 1800s. The city started burning coal in the 1820s, and after the city was chartered in 1836 it grew rapidly.[45] As early as 1860, medical professionals investigated the effects of fire-based energy and identified "local constituents of the atmosphere" as "exceedingly irrespirable and poisonous."[46] Doctors in Cleveland not only identified carbonic acid and oxygen deprivation issues with coal but also found the "influence of illuminating agents as candles and lamps" as problematic, claiming that "two candles require as much oxygen as a full grown man."[47] By the 1870s, Cleveland was becoming much like Pittsburgh—a place known for its smoke, manufactories, and tall chimneys.[48] In 1892, art collector and teacher Charles F. Olney founded the Society for the Promotion of Atmospheric Purity, which lobbied for the city to effectively enforce its antismoke laws. Olney, in a familiar theme of Victorian Progressivism, claimed that "the gospel of cleanliness can be preached and cultivated only under favorable conditions," and that the "proper aesthetic" in the city could elevate all human beings.[49]

While Cleveland dealt with smoke on the shores of Lake Erie, Chicago's issues were similar. Although Chicago had passed a city smoke ordinance in 1881 and had established a department of smoke inspection, the department's early accomplishments were questionable.[50] Evidence suggests that offending corporations either ignored fines or paid them and continued emitting smoke.[51] In 1890, for example, the Chicago Department of Health observed 3,215 violations of the city's smoke ordinances, served notices to 2,189 of those violators, and "abated" 746 of the nuisances. Two hundred and ninety-nine of the violators that year were fined $50 each.[52] As in Cleveland, Chicago's smoke problems were caused by the burning of inexpensive and accessible high-sulfur bituminous coal—mostly providing heat and steam power. Although little was known about the makeup of coal smoke at the time, many people considered it a serious problem in Chicago.[53] Both middle-class women reformers and Chicago businesses waged campaigns to assist the city's ineffective health department in the enforcement of smoke ordinances.[54]

As in Cleveland and Chicago, most major Midwestern cities that utilized bituminous coal for their primary energy source established smoke ordinances, and private groups of citizens urged enforcement.[55] City officials and citizen reformers alike did not seek to eliminate coal. The widespread belief was that smoke represented improper combustion and that properly applied technological solutions could bring the problem under control. Although John Tyndall suggested that carbon dioxide in the atmosphere might be responsible for climate change as early as 1861, and Swedish scientist Svante Arrhenius established the link between carbon dioxide from fossil fuels and climate change a few decades later, there is no indication from the historical record that these were public concerns until nearly a century later.[56] As noted by Christine Meisner Rosen in her study of smoke in Chicago in the late 1800s, little was known about invisible airborne particulate matter and greenhouse gases—it was the visible soot and tangible effects of smoke that were a public nuisance.[57] Smoke abatement in the late 1800s could more properly be termed smoke control. Unaware of the threats posed by the concealed components of smoke, Progressive Era Americans were concerned with energy's salient qualities that manifested themselves in visible soot and unpleasant air.

Responses to the smoke problem varied. Some struggled to resolve the paradox between smoke and progress, while others denied that an issue existed. While reformers attempted to abate the smoke nuisance, inventors and engineers sought technical solutions. Although natural gas, coal gas, and petroleum were in limited use by the 1870s, none of these were viable alternative energy sources in the late 1800s. Burning gases and petroleum fuels still involved proximal flame, and required distribution systems that were cost-prohibitive and not yet technically feasible.[58] For both industry and domestic use, fossil fuels made up the bulk of the nation's established energy sources, and society sought technologies to eliminate smoke, soot, fumes, and fire more than they actively sought energy alternatives. A small sampling of US patents issued in the year 1880 serves as an example. For wind-power-related inventions, including new windmill designs and improvements, at least twelve patents were granted.[59] In the same period, over fifty patents were issued for smoke-consumption devices and spark arrestors for dealing with the effects of combustion-based energy. Many more were issued for improved coal stoves, burners, and lamps.[60]

Technical solutions intended to mitigate the effects of pyrotechnic energy had been sought since Ben Franklin designed his improved draft woodstove in the 1700s. With coal specifically, the core problem was the high percentage of impurities found in the soft bituminous variety that was used in most US cities—with the exception on the East Coast, where the more "smokeless" anthracite variety was used. Considering that anthracite coal delivered to the Midwest was nearly double the price of its smokier bituminous relative, choosing it for a fuel was not a practical option for industrial or domestic users.[61] For many industrialists, it was less expensive to pay fines than it was to purchase cleaner coal or to maintain and support smoke abatement equipment.[62] Most smoke abatement "experts" in the latter half of the nineteenth century maintained that smoke could be nearly eliminated by proper firing, that proper quantities of air into a furnace would prevent smoke. More often than not, "inefficient firemen" were blamed for defeating the action of automatic stokers and other devices meant to regulate air and fuel.[63] The sheer number and types of devices and methods designed to control smoke make them difficult to categorize, but the following extract from an issue of the *American Gas Light Journal* of 1880 provides sound evidence regarding the number of technical approaches:

> The inventions and devices are almost innumerable, the patents issued are several hundred and the number rejected much greater, and many of them possess merits of their own. They include: A water drum (with leg) in front of the bridge walls and inclined

bars; deflecting arches in front and rear of a hollow iron bridge wall, perforated for the supply of air; such an arch over the bridge wall, with an arrangement of the furnace doors so as to supply the proper quantity of air; surface-draft inventions, combining the application of hot steam and atmospheric air; a perforated firebox for the incandescence of the fuel in front of the fines, with a gas chamber and arrangement for the supply of air or steam; automatic feeders of various patterns, including one which pulverizes the fuel and spreads it evenly upon the burning mass under the boilers; various applications of superheated steam, gas and hot-air chambers, for the admission of air into the furnace. All these have, as stated, merits peculiar to themselves. We incline to the belief that no plan can be perfectly worked without intelligence in the boiler room—no device will display its own complete capacity.[64]

Other techniques included the "washing" of smoke, wherein a spray of water in a chimney "clean[ed] the smoke of much of its impurities," after which the water was collected and, after a "proper treatment," used as a colorant for black paint.[65] This method removed the unpleasant blackness of coal smoke, making it more palatable. It was commonly believed that removing soot from smoke rendered it into a harmless "purified" state, no longer a threat to health or environment. Devices that separated soot from smoke were revered for their ability to perform effective purification.[66] The sheer variety and number of smoke abatement devices that were developed after the Civil War for both industrial and domestic use indicate the scope of the smoke problem. Railroads also faced a smoke issue. One publication from 1880, the *Annual Report of the American Railway Master Mechanics*, indicated that "the United States Government has, by the grant of letters patent, said that more than six hundred persons have perfected new and useful improvements in preventing the entrance of sparks, smoke, etc., to railway carriages, and that over three hundred others have perfected new and useful improvements in appliances for consuming smoke, and yet no competent authority has said that any one of these nine hundred improvements are worthy of general adoption."[67]

Of all the technological solutions that were sought to eliminate the smoke nuisance, a simple study of chimneys perhaps best reveals American attitudes about space, place, and the environment in the late 1800s. In crowded residential settings such as New York City, one solution to the smoke problem was to construct taller chimneys that would carry the smoke "to more distant localities, [where it] eventually falls to the ground and contaminates the atmosphere in proportion to the extent of its diffusion."[68] In discussing zoning for factory locations in 1874, the State Board of Health in Massachusetts praised the construction of an "immense" chimney by the New England Glass Company, which carried "metal oxides, litharge, and arsenic" to such a height in the atmosphere that "before they reach anybody they are diluted sufficiently not to be noxious."[69] An 1885 book on chimney construction from the Massachusetts Institute of Technology proudly stated that the purpose of a chimney was to "convey the noxious gasses to such a height that they shall be so intermingled with the atmosphere as not to be injurious to health."[70] A few years later, at a meeting of the American Society of Mechanical Engineers, this concept was reiterated, that "chimneys needed to be of sufficient height to get rid of obnoxious gases."[71] Numerous other publications of the time speak in terms of the necessity of chimneys to "carry away the smoke," reflecting an impulse for smoke removal.[72] This idea of removing smoke from the immediate area is indicative of early attitudes about the atmosphere's ability to absorb or tolerate the nuisance. On par with landfills and open-stream dumping, any solution that made smoke invisible—either by moving it away from the energy source or removing its blackness—seemed satisfactory.

A culture of production and consumption must by necessity be a high-energy culture, and the energy technology that the United States had adopted did not fit with Progressive ideas about cleanliness, health, purity, and pastoralism. The pursuit of technological solutions to "fix" the intrinsic issues associated with coal and pyrotechnical energy sources was more than a social response to the nuisance of smoke; it was part of an impulse to restore the imagined exceptionalism of the county's environment. Whether it was smoke abatement, spark arrestors, improved lamp and stove designs, or taller chimneys, the push for technical solutions that would eliminate the inherent problems associated with the burning of coal and other fossil fuels was indicative of the American energy dilemma that began after the Civil War. Unwilling or unable to make sacrifices of convenience, profits, or consumption, businesses and urban reformers relied on engineering and technology in attempts to solve the energy quandary they had found themselves in.[73] Through a growth in consumerism and a desire for low-cost energy, urban Americans sealed their fate by accepting the emergence of coal and other dirty energy technologies by default.[74] In turn, society sought technologies to assuage, conceal, or distance themselves from most of the visible aspects of an energy infrastructure they were dependent upon. While some of these technologies were attempts to solve the problems of smoke by eliminating it at or near the point of combustion, others simply sought to put distance between smoke and society, following a philosophy of concealment. Putting space between the effluents of energy and living places added a layer of abstraction to energy use and reflected emerging ideas that defined environmentalism in terms of quality of life.[75] In an era when ecological concerns went no further than what people could see and smell outside, moving soot, smoke, and sparks away from the individual was a sufficient solution to make energy cleaner. By the second half of the nineteenth century, the United States had become a high-energy society. By that point the world's largest consumer of coal, the dissonance inherent within desires for economic progress, social advancement, and maintenance of a providential American environment were becoming obvious. While the cultural responses to the smoke issue ranged from ignorance to denial to a drive for reform, activists and inventors sought technical solutions to render invisible the most discernable vestiges of fire-based energy. In dealing with the nuisance of smoke, American society revealed an unfettered desire to consume energy without any of the unpleasant effects of energy production.

CHAPTER 4
TECHNOLOGY AND ENERGY IN THE ABSTRACT

> Just now, in civilization, and the arts, the people of Asia are entirely behind those of Europe; those of the East of Europe behind those of the West of it; while we, here in America, think we discover, and invent, and improve, faster than any of them.
>
> Abraham Lincoln

The smoke that signified the emergence of the United States as a high-energy society in the nineteenth century was not a mystery to the Americans who experienced it. Burning coal, wood, candles, or various oils was how one illuminated, cooked, smelted, forged, or produced steam for motive power. Unlike generations that would come after, nineteenth-century Americans experienced fire up close and personal in some form on a daily basis. Whether it was a fireman on a train, a family gathering wood for the hearth, or a woman trimming the wick on a lantern, the utilization of energy was an active, participatory process. Direct flame was the era's energy source, and its management was as routine then as the wall switch is today. Fuel stocks had to be maintained, soot was an ongoing byproduct, and smoke was omnipresent, especially in rapidly industrializing urban environments. The use of fuel for heat or light required physical exertion, which, as historian Christopher Jones has pointed out, provided an ongoing awareness and a strong incentive to limit energy consumption.[1]

At the same time that the effects of smoke and soot were becoming a regressive nuisance in the latter half of the nineteenth century, new technologies and emerging technical systems had begun to change communications, transportation, and energy in ways never before imagined. Newspaper headlines declaring Morse's telegraph "the most wonderful and useful invention of the age" and the transcontinental railroad as bringing the West "one hundred years nearer to the van of practical progress" demonstrated the belief that new technologies would usher in a better future.[2] From devices that would eliminate the smoke nuisance to steam engines that were "perfect," both the media and influential leaders expressed confidence that solutions for a newly remade world were just around the corner.

Confidence in technology had a long tradition in the United States, and it grew exponentially as the nation entered the second half of the nineteenth century. From the water-driven textile mills of New England to the promotion of the so-called American system of manufacturing, Americans celebrated innovation and idolized inventors and tinkerers from Benjamin Franklin to Samuel Morse. As mechanization and an emerging interest in science-based solutions began to merge in the nineteenth century, society began to embrace an ideology derived from the traditions of Francis Bacon, that science—or, at least, invention—should effect all things possible. In the grand precentennial work *One Hundred Years' Progress of the United States*, published in 1870, the author boldly proclaims, "We have no Alexander, or Caesar, or Bonaparte, or Wellington, to shine on the stormy pages of our history, we have such names as Franklin, Whitney, Morse, and a host of others, to shed a more beneficent lustre on the story of our rise."[3]

The notion that innovation was a critical component for American progress was not a new theme, but part of a long tradition of faith in technology. Even though Thomas Jefferson had not favored English-style industrialization, he believed that through innovation the United States could keep its operations "light and flexible." As secretary of state in charge of the first United States Patent Office, Jefferson believed that improvements had an important role to play in America's future, and his principle of flexibility suggested that with the right technology the nation could enjoy the fruits of modernity while maintaining its pastoral ethos. Some six decades later, the young Congressman Abraham Lincoln espoused a similar narrative regarding the power of innovation. In his 1858–1859 lectures on "Discoveries and Inventions," Lincoln's faith in technology and the US system of fostering innovation was clear: "I have already intimated my opinion that in the world's history, certain inventions and discoveries occurred, of peculiar value, on account of their great efficiency in facilitating all other inventions and discoveries. Of these were the arts of writing and of printing—the discovery of America, and the introduction of Patent-laws."[4]

Beyond men such as Jefferson and Lincoln, even the poetic often succumbed to moments of technological enthusiasm. In his *Leaves of Grass*, first published in 1855, Walt Whitman equated inventors with others holding a prominent place in the makeup of the country: "But the genius of the United States is not best or most in its executives or legislatures, nor in its ambassadors or authors or colleges or churches or parlors, nor even in its newspapers or inventors, but always most in the common people."[5] Ralph Waldo Emerson, expressing his thoughts about emerging technologies in *Works and Days*, recognized that "life seems almost made over new."[6] On the mechanical arts, Emerson continued, "These arts open great gates of a future, promising to make the world plastic and to lift human life out of its beggary to a god-like ease and power."[7]

Emerson's confidence in the promise of the mechanical arts was indicative of a deep-rooted American belief that a better tomorrow was lurking in the workshops just around the corner. Cultural historian Leonard Neufeldt observes that while Emerson described the promise of the mechanical arts, he also expressed his belief that "we must look deeper for our salvation than to steam, photographs, balloons, or astronomy."[8] Within this contradiction, where faith in technology is juxtaposed with the thought that technology could not be a true savior, a useful metaphor is exposed for society both then and now. Since the first European colonists arrived on the shores of New England, conflicting desires for both technological advancement and a pastoral city on the hill have been the American paradox. Perceptions of unspoiled American landscapes were uncomfortably dissonant with a smoky, soot-filled foreground, and society would always be hopeful that technical solutions could create a clear, progressive utopia.

In an era of great technological change, it was understandable that some would believe most or even all problems could be solved with the application of scientific principles and American ingenuity. As early as 1829, Harvard professor Jacob Bigelow spelled out in *Elements of Technology* the goal that technology could acquire dominion over nature and create a man-made environment.[9] Attempts to control the human-made world had been well under way prior to Bigelow's proclamation, but they accelerated in nineteenth-century America. Between Morse's invention of the telegraph in the 1840s, newly emerging municipal central water supply systems, and the completion of the transcontinental railroad in 1869, there seemed to be no end to the possibilities of what the future could bring. Adding a nationalist tone to American innovation, authors such as Horace Greeley tied together themes of mechanization and democratization with his ideas about "appliances of a higher life."[10] In the second half of the nineteenth century, technology and invention were celebrated in books, articles, and paintings, and many felt that invention and progress went hand in hand.

With a strong belief in progress through technology, the United States rushed headlong into an imagined modernity. As the country's mantra of manifest destiny drove expansion westward, steam-driven mechanization smoked and clanked into a period of rapid industrialization and urbanization. The nation's changes were not lost on historians. Perry Miller, writing in the 1960s, was one of the first to identify the multiplication of "gadgets" in the era as a cause of national "bemusement." Miller traced the origin of this bemusement to a reaction against European pretensions of a "theoretical superiority" implying that Americans "flung themselves into the technological torrent" due to "base standards of value," thus tying technological faith to capitalism.[11] When Leo Marx later picked up on Miller's writings about the notion of "technological sublime," he expanded upon the work to show how Americans came to mesh industrialization with an American pastoral ethos with its origins in the Jeffersonian notion of a middle landscape between nature and husbandry. In Marx's view, Americans directed their awe and reverence that was once reserved for the Deity to technological conquest—a thread that would become more obvious in 1876 as the rhetoric related to new technologies would soon reveal.[12] Other writers, such as Christopher Lasch and David Noble, later identified the same theme of faith in technology as a core component in the development of American society.[13] David Nye, echoing Marx's assertion, defined the respect for technological advancement as "a preferred American trope" and a "religious feeling aroused by confrontation with impressive objects."[14]

As the nation reached its centennial year, enthusiasm for technology had accelerated exponentially. New inventions such as Alexander Graham Bell's telephone and the first electrical systems were introduced at the Centennial International Exhibition in Philadelphia in 1876, the nation's first world's fair. Faith in new devices and systems combined with strong themes of nationalism that contributed to a general acceptance of all things technical. As advancements were praised they took on a new life that became impervious to question as confidence in technology trumped concerns over any possible consequences.

In the midst of American's fascination with gadgetry and mechanization, the nation's centennial celebration became a launching point for a new paradigm, the accelerated emergence of technical systems. Whereas transportation, communication, and water utility networks were beginning to be established prior to 1876, new technologies that utilized interconnections as a core component of their operational processes were becoming more common. By their very nature, technical systems altered the continuity of processes and affected social perceptions of their components, which in turn took on new meanings. Newly emerging systems began to partition or render their components autonomously; arc lights would be seen as stand-alone devices for illumination, while the connecting wires and dynamos that generated power would fade into the background. According to historian Kevin Borg, systems tend-ed to render processes "ontologically opaque" as they separated points of origin from points of consumption.[15]

The technological systems that first appeared in the mid-1800s began to alter spatial dynamics via tracks, wires, and pipelines. Morse's telegraph in the 1840s established networks that would alter perceptions of communications, and the expanding railroad transformed both shipping of goods and passenger transport. For the first time ever, the transmission of messages involved intangible, invisible processes via the telegraph key that separated sender and receiver. The first municipal water systems, via invisible underground networks, brought fresher, cleaner water from distant sources into homes inconspicuously. Railroad networks, though quite visible, began to obscure the space between the farmer and the consumer and in the process began to transform the social consciousness of systems of production.

William Cronon addresses the dynamics of transportation systems in the context of space, or separation between commodity production and consumption as it applied to agricultural commodities. In *Nature's Metropolis: Chicago and the Great West*, Cronon investigates how technological developments altered the perception of consumables such as grain and meat. Using Chicago as an example, Cronon follows the development of the grain mill and the meat packing industry. For Cronon, the transformation of grain into flour, and a steer into packaged beef, "partitioned a natural material into a multitude of standardized commodities."[16] In Cronon's analysis, the mode of transmission that was transformational was the railroad track and railroad car—the ability to transport physical, tangible goods severed the links between the commodities of the economy and the resources from the ecosystem. In this scenario, the living animal became a commoditized package of dressed meat and the field of grain became bags of white flour. When looking at communication and energy technologies as well as emerging infrastructure systems such as municipal water supplies under a similar lens, transmission systems consisting of wires and pipelines become equivalent to Cronon's railroad—technological pathways that abstracted processes or materials. As with packaged beef, the link between person and telegram, power generation and consumption, and reservoir and faucet became obscured. The transformation in perceptions due to the advent of newly emerging technical networks was subtle yet pervasive as processes and goods took on new meanings in the second half of the nineteenth century. Long-embedded paradigms of physical relationships became altered as infrastructures transformed the dynamics of distance, and technologies became redefined.

The impact of emerging technical systems reached a turning point at the Centennial Exhibition. As a historical milepost, the event has been exhaustively covered by historians in terms ranging from that of a "moral influence" to an "industrial revelation to the rest of the world."[17] When interpreted through the lens of the history of technology in general and more specifically the history of technological systems, the exhibition of 1876 takes on new meanings. As a showplace of new technologies, the fair serves as a case study in how social perceptions can become altered through the application of systems. Suddenly, fire, coal, and steam began to be hidden, networks became opaque, and the contrivances on display took on a perception of independence, cognitively separated from components they relied upon. Not only would the fair demonstrate the perceptual dynamics of emerging sociotechnical systems, it would also serve as a point in time when new technical paradigms would become embedded.

Perhaps no single engine in history has been written about as much as the Corliss steam engine that was the centerpiece of the Philadelphia exhibition. As the main attraction of the fair's indoor pavilion known as Machinery Hall, the engine has been at the center of lengthy narratives by social historians who have found deep symbolism in the celebrated machine. At more than fifty feet tall with a thirty-foot-diameter flywheel, the twin-cylinder behemoth tacitly powered an assortment of mechanical devices spread throughout the hall. Although the Corliss ran on coal-derived steam, few if any accounts of the engine at the time or since mention the words *steam* or *coal* in connection with the exhibit. The hidden system of underground piping, boilers, and burning coal became consciously disconnected and inconsequential in the presence of the virtually silent machine. The Corliss engine was just that—the Corliss; it was not referred to as a steam engine or a coal engine, but an entity that was an autonomous actor, a thing of beauty and technological sophistication that was perceived apart from its sustaining infrastructure. In the historical record, the Corliss serves as a lesson in how system components are construed based on context.

While fairgoers and journalists were praising the clean, silent grandeur of the refined engine inside Machinery Hall, outside the Philadelphia fairgrounds the smoke and noise from the first experimental steam-powered streetcars were reported as a nuisance, and boiler explosions were a regular feature of the daily news.[18] The contrast between the steam-powered streetcars and the Corliss is indicative of how technology is conceptualized when it is separated from its operational network. The detachment of infrastructure from operation may not have started with the Corliss, but the narrative surrounding the great engine is demonstrative of how technologies can take on new meanings as components of obscured systems.

William Dean Howells, the influential editor of the *Atlantic Monthly*, referred to the Corliss as an "athlete of steel and iron, with not a superfluous ounce of metal," and marveled at the pistons thrusting in a "vast and silent grandeur."[19] Another enthusiastic author said that the Corliss was "true evidence of man's creative power; . . . Prometheus unbound."[20] California poet Joaquin Miller's view of the machinery was that "it is the acorn from which shall grow the wide-spreading oak of a century's growth."[21] When poet Walt Whitman entered the building in his wheelchair, he asked his escort to stop in front of the Corliss, where he remained transfixed for nearly thirty minutes.[22] In its silence and separation from the coal and fire that ultimately powered it, the Corliss became a gentrified independent actor. The process of energy production and utilization suddenly became less visible as the coal-burning boiler house was isolated from the motive power of the Corliss. The engineering savvy and vast array of new gadgets in the hall captured the attention of attendees and historians alike, but the thirty tons of coal consumed per day to run it all was far removed from the minds of those in awe.[23]

On May 10, the opening day of the exhibition, an article appeared in the *New York Times* with the headline "The Yost Murder Trial: What Detective M'parlan Knows about the Molly Maguires' Secrets." An alleged secret organization of Irish Catholic coal miners, the Molly Maguires, mined the anthracite coal that ran the fair and were blamed for a series labor strikes and violence beginning in 1875. When coal owners and operators cut their wages, the public initially supported the miners in their protests—until Franklin B. Gowen, mine operator and union buster, swayed public opinion. Gowen waged a campaign to frame the workers, bringing five miners to trial over the suspicious killing of police officer Benjamin Yost. Several miners were quickly convicted and eventually hanged.[24] Beneath the Yost murder story in the same issue of the *Times*, a story titled "The Centennial Exhibition" promoted the opening of the fair, an event noted as "an institution whose spirit and purpose belong emphatically to the present age." The pageantry and technological allure of the Corliss not only obscured the burning of the coal that ultimately powered it but also broke the link between its primary fuel source and the workers who produced it. While the Corliss was praised as an engineering marvel by those who saw it, it represented the end point of a long supply system of the coal production that fed it. In the process of transforming coal to fire, fire to steam, and steam to motion, the digging of the coal and the exploitation of the workers who dug it were forgotten by those who marveled over the engine. The Corliss could not have started without the coal that powered it, and the coal would not be available without the Irish American coal miners who mined it. Advanced energy technologies had already begun to blur the direct link between fire and energy, and they also obscured connections up and down the primary supply chain, rendering both labor and environmental effects inconsequential.

While the steam engine itself had been around for over a century, the Corliss in Philadelphia began to change the way those in attendance perceived motive power, as it was a machine of fascination at one end of a small technical system. The one-fifth of the US population who saw the Corliss up close—and for many it was likely their first close encounter with steam-driven machinery of this scale—saw only the manifestation of clean and silent energy. Invisible from the environs of the hall were the processes of mining, transporting, and burning of the anthracite coal that supplied its power.[25] Depending on the precise mine where the coal came from, it traveled via a coal-driven train a few hundred miles from the anthracite regions of Pennsylvania, and onto the fairgrounds at Fairview Park in Philadelphia. Once there, the coal was dumped into chutes in the boiler house annex No. 2, where each day it heated six hundred thousand gallons of water in twenty boilers.[26] Connected to Machinery Hall by 320 feet of double-riveted 18-inch diameter piping, steam annex No. 2 supplied what engineers referred to as "life power" to the giant Corliss.[27] The exhaust from the burning coal in the boiler house exited into the atmosphere via two 91-foot-tall decorative brick chimneys. Once the superheated steam was delivered to the Corliss and converted to mechanical motion through the two massive cylinders, the rotary energy was transferred to a variety of machines in the fourteen-acre hall by way of 10,400 feet of shafting.[28] Through a complex transmission system of pulleys and belts, the Corliss brought life to dependent devices throughout the building. By the time the coal-derived mechanical energy was rendered into useful power, its origins were completely detached from the infrastructure on which it relied. Those who saw the Corliss and the equipment it powered did not see the Molly Maguires, nor the men shoveling coal into boiler annex No. 2, nor did they experience any of the smoke or fumes emitted by the deceptively embellished chimneys.

On its own, the Corliss in Philadelphia did not magically change the way society perceived steam technology—smoking locomotives, exploding boilers, noise, and other issues associated with steam power were demonized both before and after the 1876 fair. It was not uncommon to read of steam engines and associated boilers being referred to as a menace or a nuisance. Typical of the day were articles that associated steam engines with "nervous strain," yet no such terms are used in accounts that describe the revered Corliss.[29] Instead, the rhetoric associated with the Corliss at the fair demonstrates how society saw technology when its most negative characteristics were obscured. Machinery Hall was a microcosm that showed how systems could change the social meanings attached to technology, not only by breaking the connection between miners and coal but also by breaking the ecological connection between mining, smoke, and carbon emissions. Although exploitation of natural resources and air pollution were not social concerns in 1876, the presentation and perception of the Corliss as a piece of technology disconnected from coal, fire, and steam is an important point when considering how the past informs the future. The Corliss directly promoted the idea that steam engines and coal power could be rendered as clean technologies.

The Corliss represented the most refined manifestation of steam power to date. In its polished and larger-than-life form, the engine was, as science historian George Basalla later noted, a central symbol of American progress and part of one of the first energy myths to sweep the nation.[30] Coal was seen by some as God's gift to the United States, and although smoke and soot were encroaching byproducts of industrialization, the Corliss gave the impression that American ingenuity could overcome all the negatives associated with the past. While the Corliss is an example of how technological enthusiasm and newly emerging systems could alter perceptions, it also shows how new contrivances and systems inspired a spiritual reverence in a modernizing society. The Corliss was seen as a providential technology, an American steam engine that transcended the rhetoric of nationalism.

In a sermon titled "The Present Condition, Prospects, Beneficent Work, Needs and Obligations," Presbyterian pastor Reverend Edward Humphrey equated the Corliss to force, force to God, and steam to Christianity:

> The Corliss engine furnishes the positivist with a fresh illustration of force. 'Here,' he exclaims, 'is force indeed—force all but

irresistible. This at last solves the problem of the great first cause: Force is God.' ... Now a more thorough examination of the Exhibition will show that Christianity is the parent of the forces which are changing the face of the world . . . The steam-engine, the steamship, the locomotive, the railway, the magnetic telegraph, are products of Christendom.[31]

The American Board of Commissioners for Foreign Missions echoed Humphrey's sentiments:

> Like the city of mechanism, a department in the centennial Exhibition in Philadelphia, silent and moveless around the grand Corliss engine, also silent and moveless, hand, wheel, and cog, all adjusted, waiting for a single touch to a single spring or lever to start the whole to one vast, simultaneous, mighty life, so now the world, with its preparation, seems waiting the access of the Spirit's power promised to the prayers of the people of God.[32]

Ironically, at the same time that the Corliss had reached a spiritual status by representing the most advanced steam technology of the day, its allure was about to be usurped by a group of inventors in the same building. As a case study, the Corliss demonstrates how the combination of enthusiasm for technology and innocuous systems could shape public perceptions. The Corliss as it stood in Machinery Hall tamed coal-derived steam power through a system-derived sleight of hand that kept the most undesirable characteristics of its operation out of view. Despite the grand reception of the Corliss, the public fascination with steam technology would both peak and begin to fade at the same time.

On June 25, 1876, a crowd of dignitaries had gathered in Machinery Hall at Alexander Graham Bell's exhibit for a demonstration of his speaking telephone.[33] Among the guests were William Wallace, one of the inventors of the Wallace-Farmer electrical dynamo; the notable Irish physicist Lord Kelvin, who was at that time Sir William Thomson; and Dom Pedro II, Emperor of Brazil. As Sir William grew impatient for the event to begin, Bell rushed making last-minute adjustments preparing to demonstrate his new device. Sabbatarianism had closed the fair to the general public for the day, leaving perfectly quiet conditions for the demonstration, as Bell had requested. With wires strung through the hall, nearly one hundred yards separated two of Bell's single-pole transmitters. With Sir William on one end, and Dom Pedro on the other, Sir William spoke into the transmitter, "Do you hear what I say?" at which time Dom Pedro leaped from his chair exclaiming, "I hear, I hear!"[34] The emperor's reaction, as well as that of Sir William, was indicative of the excitement brought about by the new electric-based inventions and processes. Through a variety of technologies, wires and electrons became a new channel of abstraction through which invisible processes moved. In anticipation of broken barriers of space, Sir William later wrote of the experience:

> I need scarcely say I was astonished and delighted; so were others, including some judges of our group who witnessed the experiments and verified with their own ears the electric transmission of speech. This, perhaps, the greatest marvel hitherto achieved by the electric telegraph, has been obtained by appliances of quite a homespun and rudimentary character. With somewhat more advanced plans and more powerful apparatus, we may confidently expect that Mr. Bell will give us the means of making voice and spoken words audible through the electric wire to an ear hundreds of miles distant.[35]

While an undulating current generated by the phone itself powered Bell's telephone, the transmission of energy—electrical current through wire—was on display at the fair and was the harbinger of things to come. The Centennial Exhibition was not the launching point for electricity or wire-based communications, but it was where many Americans experienced the invisible force for the first time.

Two major displays of dynamo-generated electrical power were present in Philadelphia: the Wallace-Farmer and the Gramme exhibits. The Wallace-Farmer dynamo operated large arc lights for night illumination at the fair, one of the first installations of its kind, and the Gramme exhibit powered not only arc lamps but also an electroplating device and a motor connected to a pump that raised water for a small waterfall.[36] Both the Wallace-Farmer and the Gramme dynamos were rotated via pulley, ultimately driven by steam derived from coal. For Dr. Joseph Henry, a distinguished professor and judge at the fair, the use of coal "combustion" to drive the dynamos represented a "transmutation of energy," which he felt was a significant advancement or improvement.[37] Henry's term *transmutation*, based on the word *mutation*, was an apt description for the processes in place to generate electricity. The dynamos were part of a system that moved both coal and steam out of view as the pent-up energy in coal was converted first to heat, then to steam, then to mechanical energy, and finally to electricity. The Wallace-Farmer and Gramme exhibits were some of the earliest public demonstrations of secondary energy, or energy derived from a primary energy source.

As spatial dynamics and system visibility played a central role in how technologies abstracted process, other characteristics also contributed to the perceptual turn that separated technology from nature. Qualities such as appearance, precision, and speed contributed to a growing faith in technology that accelerated in 1876 as the nation suddenly saw what was possible—the past was the pony express, the packet ship, the lantern, and steam; the future was the dynamo, the telephone, and the electric light. The past was stacking wood, shoveling coal, and pouring oil; the future was the promise of new forms of energy that removed direct human interaction with fire and fuel. Enthusiasm over technology began to suppress any future concerns over consequentiality and detached people from process and process from nature.

Just as the Corliss took on an independent identity separate from its ultimate source of power, electricity was about to become an autonomous entity as well. Elihu Thomson, who would eventually establish the Thomson-Houston Company, the forerunner of the General Electric Corporation, recalled the Centennial Exhibition as the launching point for the technology of electricity: "No similar period in the world's history has in any art shown so rapid development, so extensive and refined scientific study and experiment, so active invention, so varied application, such care and perfection in manufacture, as has taken place in the electrical field since the Centennial Exhibition."[38] In addition to the Gramme and Wallace-Farmer exhibits, a variety of other electric devices were on display, including several battery-powered motors.[39] While the battery-powered technologies were embodiments of electricity in action, the dynamos were most relevant to the concept of abstraction, because they were ultimately powered by coal-derived steam and were the precursors to modern electrical systems. In these iterations of generated energy, power was not just "transmuted," to use Henry's word, but was converted from visible power, which involved a transition from fire and smoke to the rotary action of shafts and belts, to invisible energy, which added a deeper layer of abstraction. As with the system of coal hoppers, boilers, and underground steam lines that fed the Corliss, visibility and distance between generation and consumption had once again been altered. Appropriately, Wallace had called his generator a "telemachon" (*tele* from Greek, meaning "distant") and stressed that its purpose was for transmitting useful electric power over distances. It was the same Wallace-Farmer dynamo that two years later piqued the interest of Thomas Edison, who developed his own system of power generation and lighting. Edison was struck by the dynamos and was anxious to acquire the generators, writing repeatedly to Wallace to "hurry up" on the shipment of a dynamo to Edison's Menlo Park laboratory in 1878.[40]

Although Americans' old conceptions of space had been fracturing since the development of the telegraph and the build-out of the railroad, especially the first transcontinental in 1869, the technologies on display at the Centennial Exhibition further shattered the boundaries of the past. Distance and invisibility removed human interaction from familiar processes and began to create a perceptual sphere that transcended nature. Suddenly, Bell's telephone, the various displays of electrical energy, the Corliss engine, and other contrivances on display were not only technical improvements and advancements but autonomous technologies that represented a mastery over nature. This was especially evident in processes that utilized wire and harnessed the invisible force of electricity.

The technologies of wire that the American population saw in 1876 represented a paradigm shift of unprecedented proportions. Prior to electricity, all energy had been derived from visible primary sources. Heat and light, whether in the household or commercially, involved visible flame along with the byproducts of combustion. Waterwheels and steam engines involved moving parts such as shafts, pulleys, and gears that were experiential and comprehensible. Locomotives, as the most common manifestations of steam engines for Americans at the time of the centennial, were loud contrivances that emitted sparks, smoke, and soot. Although the Corliss would go far to tame steam and coal in Machinery Hall, the technologies of wire had the biggest impact on American society moving forward. Electricity was at once modern, unseen, and mysterious; it was, according to *Webster's Dictionary* at the time of the fair, "a power in nature, a manifestation of energy," and not tangible.[41]

Even though the Corliss was the center of attention, enthusiasm for all things electrical ran high in Philadelphia in during the centennial year. On July 2, three miles from the exhibition at the National Centennial Commemoration, speakers on the steps of Independence Hall gloated over the accomplishments of the nation. The influential Leverett Saltonstall, Massachusetts State Commissioner to the Centennial Exhibition, gave "electric wires" a central role in the transmission of American intelligence around the globe:

> Huge factories of every kind greeting the eye of the traveler, and the hum of varied industry filling his ear. Great steamships lie at anchor in the harbors, and the yellow harvest falls before the march of the reaping machine. The electric wires, thanks to Boston's son but Philadelphia's patriot sage, are transmitting intelligence quicker than the lightning's flash from one side of the continent to the other, and even under the ocean to continents beyond.[42]

An analysis of language used during the Centennial reveals a prevalence of terms related to electricity suddenly in use. Along with Saltonstall, Patrick Henry's grandson, William Hirt Henry, said that it was "the electric spark emitted by the genius of James Otis [that had] kindled the genius of John Adams."[43] He went on to say that the "pent-up indignation of America" burst forth as by "an electric flash suddenly discharged" in the form of resistance to the Stamp Act.[44] During a rather long-winded speech following Henry, the Honorable Francis Putnam Stevens, a state senator from Maryland, managed to weave together themes of American canals, the first steam passenger railway, the telegraph wire, George Washington, and the Star-Spangled Banner, all in a single paragraph:

> Maryland, through her representative, Thomas Johnson, Jr., nominated in those halls George Washington to be Commander-in-Chief of "the armies raised and to be raised," and to-day her monumental city points with just pride to the noble shaft she alone has reared to his memory. She sent here such representatives as Samuel Chase, Charles Carroll of Carrollton, John Henry, Jr., Thomas Johnson, Jr., Thomas Stone, and William Paca. She boasts "The old Maryland line," with Howard, Williams, Gist, Smallwood, and others; she gave you the first telegraphic wire, the first canal, and the first steam passenger railway. It was Maryland that gave to you, and to the world, your national anthem, "The Star-Spangled Banner.[45]

A jubilant writer for Harvard's *Friends' Intelligencer*, celebrating the "unity of the English-speaking world," explained that as "Centennial bells ring out . . . peace and goodwill to all men [and] as the electric cable joins the lands, so it should join the hands, and with them the hearts; once unite those who speak the old language whether from the Old Country, from the States, from the Dominion, or from Australasia, and the peace of the world is secured."[46] Coincidentally, the same issue of the *Intelligencer* includes an essay titled "Save The Forests," bemoaning deforestation in Pennsylvania and explaining that the "temperature of a country, as well as its annual rainfall, is affected by the existence or destruction of its forests."[47] The presence of these two threads twenty pages apart in the same publication is evidence of the perceptual split that was already occurring between technology and nature—that technological advancement was good, and that ecological destruction was bad, but the two were mutually exclusive.

Those who praised the various innovations at the fair frequently used language such as "superior," "curiosity," and "force" but when it came to wires and anything electrical, commentators elevated their praise to an ethereal activist religious tone. For Christian missionaries, electric lines were poised to be the conduits that would drive the global shift to Christianity:

> All these aspects of the times evidently look toward some great event in the kingdom of God, in the not distant future. For it the whole creation waits. It is a period of vast preparation and expectancy, like the halt-hour's Apocalyptic pause in heaven on the opening of the seventh seal. Preparation and arrangement for some continuous, simultaneous impulse through the earth seem well nigh completed. The lines laid, connections and combinations established, the chain work of electric conduction complete in its links, there waits only the celestial flash, the fire from heaven.[48]

In her essay "A Centennial Outlook," a Mrs. M. B. Norton tied together gender roles, Christianity, and nationalism as she asserted that the centennial celebration marked the beginning of a turning point for Christian women. Now that the "iron bands which span [the] continent are filled with electric currents that thrill the air," and "telegraph lines" were in easy reach, Christian women would become "largely emancipated from domestic drudgery."[49]

Along with the mechanical contrivances at the fair, electrical technologies were also seen in terms of nationalism, religion, and morality by many of those who attended, yet the invisible forces within wires rose to a higher level of sublimity. Abstracted beyond the earthly bounds of mechanization, electrical forces were transformational and completely detached from the physical infrastructure that generated them. Suddenly, electricity became embedded in the minds of those who witnessed it as independent, miraculous, and utopian, yet far removed from this imagined utopia were the pits in the earth, the boiler rooms, the fires, the smoke, the steam engines, and the dynamos. Suddenly, the chain of separation between the generation and consumption of energy became longer and less visible, and as a result any future environmental or social consequentiality would become less visible as well.

While electricity began its march to become an embedded energy technology in Philadelphia in 1876, so did its method for generation. The Gramme and Wallace-Farmer dynamos ultimately depended upon coal and steam to produce their charge, and despite the smoke-free fairgrounds, the modernity of the telephone, and the smooth and silent grandeur of the Corliss, the entire exhibition was dependent upon burning coal. Alternative energy sources were not viable or scalable options, and as with the inventions of John Etzler and those who had come before, noncoal alternatives in 1876 were met with skepticism and dismissed.

As coal and the Corliss monopolized Machinery Hall, they did so at the expense of presenting the possibility of alternative energy sources. John Ericsson, the man who conceived of and built the ironclad *Monitor* of Civil War fame, was furious at the Centennial Exhibition committee's failure to invite him to the exhibition to show his solar-powered "caloric" engine.[50] Ericsson's omission as an invited guest is curious considering that he was one of the best-known engineers in the country at the time. Whether or not the coal interests influenced the decision of the centennial board to keep Ericsson's solar engine out of the fair is unclear, but unspecified members of the "coal trade" were large contributors to the Centennial committee, pledging $25,000 to support the event.[51] Ericsson responded to the fair board's refusal to allow him to display his work by publishing a 664-page book entitled *Contributions to the Centennial Exhibition*. In the preface he explains:

> The Commissioners of the Centennial Exhibition having omitted to invite me to exhibit the results of my labors connected with mechanics and physics, a gap in their record of material progress exceeding one-third of a century has been occasioned. I have therefore deemed it proper to publish a statement of my principal labors during the last third of the century, the achievements

of which the promoters of the Centennial Exhibition have called upon the civilized world to recognize.[52]

Ericsson went on, some five hundred pages into his work, to present his calculations for deploying solar engines that would utilize a solar heat strip in the west:

> Computations of the solar energy wasted on the vast areas thus specified would present an inconceivably great amount of dynamic force. Let us, therefore, merely estimate the mechanical power that would result from utilizing the solar heat on a strip of land a single mile in width, along the rainless western coast of America. ... Adopting the stated length and a width of one mile as a basis for computation, it will be seen that this very narrow belt covers 223,000 millions of square feet. Dividing the latter amount by the area of 100 square feet necessary to produce one horse-power, we learn that 22,300,000 solar engines, each of 100 horse-power, could be kept in constant operation, nine hours a day, by utilizing only that heat which is now wasted on the assumed small fraction of land extending along some of the water-fronts of the sunburnt regions of the earth. Due consideration cannot fail to convince us that the rapid exhaustion of the European coal-fields will soon cause great changes with reference to international relations, in favor of those countries which are in possession of continuous sun-power.[53]

Despite the fact that Ericsson sent a copy of the book to each judge at the exhibition, his work on solar power was ignored in 1876 while the praise of the Corliss and a number of other American technologies enthusiastically continued.

The adoption of coal-driven dynamos was a technological paradigm with a long reach. By refusing the entry of Ericsson's solar technology to the fair, the exhibition committee also denied alternative energy a broader audience. Edison, George Westinghouse, Elihu Thompson, Charles Brush, and others who became instrumental in developing systems for electrification never had a chance to witness Ericsson's work in Philadelphia, but they did witness the excitement and enthusiasm associated with invisible power and coal-driven dynamos.

From a standpoint of scalability, Ericsson's solar technology was not a practical option to replace coal, yet its presence in the fair might have led to the dissemination of his concepts. Ericsson knew his engine was not yet viable, but he was also aware of its potential, writing in 1869:

> It is true that the solar heat is often prevented from reaching the earth. On the other hand, the skilful engineer knows many ways of laying up a supply when the sky is clear and that great storehouse is opened where the fuel may be obtained free of cost and transportation. At the same time a great portion of our planet enjoys perpetual sunshine. The field, therefore, awaiting the application of the solar engine is almost beyond computation, while the source of its power is boundless.[54]

Despite Ericsson's prescience, in 1876 there was not yet an urgent need to consider alternatives to coal. Smoke was a nuisance, but the impetus at the time was to eliminate the byproducts of combustion through controlled burning, taller chimneys, or smoke control devices. If one accepts the view held by Basalla, that continuity plays an important role in the evolution of technology, Ericsson's engine represented a discontinuity that had not yet reached its potential.[55]

The degree of technological abstraction driven by spatial dynamics and faith in technology that gained momentum in 1876 ran deep. Not only did Americans see new devices and technologies at the fair, they witnessed the beginnings of a shift in energy forms and modes of delivery that would remain in place to the present day. The distraction and fascination with machines such as the Corliss and the technologies of wire on display in 1876 marked a point at which energy started to become abstract. The effect of this abstraction became pervasive as even contemporary historical accounts of the fair recount that "The Philadelphia Centennial Exhibition of 1876 was the last great Exhibition based on steam power."[56] While it is true that the mechanical energy at the fair was derived from steam, the steam was derived from coal. After 1876 all other fairs ran on steam as well, only in these later cases the steam was converted to electricity. Through the rhetoric we see how steam and coal fade into the historical background and electricity becomes construed as an independent source of energy. On the visible level, the exhibition did run on steam power, but without coal it would not have run at all. Had each guest had to first dig for and then shovel coal into a boiler while experiencing the flame and smoke prior to watching engines run and lights illuminate, the social meaning of technology in 1876 might have changed, but in Philadelphia, coal began to be physically detached from energy in the American mind. Seemingly magical, electricity became a sovereign energy source that impressed all who experienced it at the fair. A force long seen as mysterious had finally been harnessed, and an illusory clean future was just around the corner.

CHAPTER 5
OF FLUIDS, FIELDS, AND WIZARDS

> Electricity in its ordinary every-day uses surpasses all the feats of the ancient magi or modern prestidigitators. Sending light, heat, power, signals, and speech to a distance over a wire, the phenomena of induction, the transfer of metals as in electro-metallurgy, and the numerous other uses to which electricity is applied in the arts, are all truly magical and mysterious, since the most profound is unable to assert the true nature of this subtle force.
>
> "Electrical Magic," *Scientific American*, **1881**

The electrical technologies on display at the Centennial Exhibition in Philadelphia in 1876 represented a new phase in the development of practical applications for electricity. While the telegraph had been in service for nearly three decades, more electrical devices, along with power generation systems, began to appear in Philadelphia. Lights, motors, and other contrivances were powered by invisible electricity, coursing silently through wire, which continued to alter the spatial dynamics of energy and communications, departing from energy paradigms of the past. While the fair marked the debut of technologies that abstracted energy by blurring the line between power generation and consumption, it was also a point where electricity as a natural force continued to transition from a theoretical to an applied entity.

The advances that led to the technologies on display in 1876 had begun nearly a century earlier by natural philosophers who discovered how to produce and control electrical current. As an invisible force with properties unlike any encountered before, electricity was difficult for both early theorists and lay members of society to define, and to many it remained a mystery. As research into electricity transitioned from a theoretical to an applied art, discoveries of how to generate and transport current led to the technologies on display in Philadelphia. The earliest applications of applied electrical power included Samuel Morse's telegraph and Alexander Graham Bell's telephone, and although these did not represent a technology that filled a universal social need at the time, the incandescent light and first electrical systems developed after 1876 did. The promise of an affordable technology that could supplant open flames for illumination indoors was highly anticipated and added to the growing perception that a new electrical age was at hand. While coal smoke continued to be a problem for the outdoor air in most major cities, society's encounter with smoke-free and flame-free illumination in the late 1800s led to divergent perceptions in which coal, fire, and steam represented the past and electricity represented the future. Born of Enlightenment science and adopted by an America open to progressive technology, the nebulous force of electricity quickly became disassociated from coal as the variety of its social applications grew.

While many critical advancements in electrical technology before 1876 came from Europe, Benjamin Franklin made significant contributions prior to the American Revolution. In an early instance of international open-source collaboration between 1740 and 1800, Franklin corresponded with a network of Europeans sharing insights into electrical principles during a period of time that some historians refer to as "the first age of electricity."[1] Franklin's well-historicized experiment utilizing a kite and key to prove that lightning was electricity, although questioned by some, became deeply embedded in American folklore—a cultural nuance claiming the discovery of electricity as occurring on US soil.[2] Franklin recognized the possibilities that the mysterious force could bring, and was credited by his contemporaries for realizing that electricity was "an invisible, subtle matter, disseminated through all nature in various proportions, equally unobserved, and . . . all those bodies to which it peculiarly adheres are alike charged with it."[3] Franklin's observations inform nearly all characterizations of electricity from that time on, as something peculiar and difficult to conceptualize. Franklin envisioned the force as an "electrical fluid," a useful analogy to describe a phenomenon that was largely unknown and challenging to define.[4] Working with static electricity and Leyden jars, Franklin's most significant theoretical achievement was his work on the theory of electrostatic induction, which led to insights about symmetries between positive and negative charges and their simultaneous production. His discoveries about electrical charges always occurring of equal and opposite magnitude are known to this day as the Franklinian principle of conservation of charge.[5]

Although several of Franklin's theories were later disproven, his experiments and data influenced Europeans such as Alessandro Volta and Luigi Galvani, whose work at the turn of the eighteenth century went far to advance theories about electrical current. In one of the most comprehensive and respected works that trace the history of electricity, *Electricity in the 17th and 18th Centuries: A Study of Early Modern Physics*, historian John L. Heilbron points to the importance of Franklin's early work on electrostatic energy as a stimulus to further research in the 1700s.[6] For example, Galvani's famous experiments with electricity and frogs' legs in the 1790s led him to conclude that such a thing as "animal electricity" was an internal phenomenon and that an electrical charge resided in nerves and muscles of the body.[7] Later, Mary Shelley exploited the conceptualization of electricity as a metaphysical force in her novel *Frankenstein*, in which all that is needed to bring a being to life is a vigorous jolt. While Galvani was studying at the University of Bologna, Volta was conducting research just to the west, at the University of Pavia. Upon reading Galvani's work, Volta began to conduct his own experiments, coming to a conclusion opposite of that of Galvani—that electricity was not an internal force of the body, but an external force that could be created when two dissimilar metals came into contact. The theory that electricity could be created by means other than a static spark, was perhaps one of the most significant discoveries in modern history, as it was the beginning of a new branch of electrical discovery based on current instead of static.

By the late 1700s, the science of electricity had begun to take a momentous turn as Volta went on to invent the "pile" in 1800, which was in essence the first practical battery.[8] Volta's pile marks a point in time when electricity could be summoned on demand and controlled, a significant step on the way to electricity as a force that could be harnessed. As posited by a number of theorists including Heilbron, Volta's pile and the steady current it provided later became a critical link in establishing the relationship between electricity and magnetism, opening up a yet another branch of electrical research: electromagnetism.[9] Utilizing the pile that Volta perfected, Hans Christian Ørsted at the University of Copenhagen discovered the vital link between electrical current and magnetism in 1820.[10] Ørsted achieved his breakthrough as he observed a compass needle moving when it approached a wire carrying electrical current derived from a voltaic pile.[11] Not only did this simple observation prove a link between electrical force and magnetism, but the moving needle also demonstrated how mechanical motion could result from the interaction of a magnet and current. In addition to the significance of Ørsted's discovery of electromagnetism was his embrace of Kantian metaphysics. Understanding the properties of the nontangible entity that is electricity required a leap into the realm of the metaphysical world, which would confound society's conceptualization of electricity. In Ørsted's day, the dominant philosophy among scientists in Europe was bound up in epistemological theories based on empiricism, yet Ørsted was led to perform his experiments by inferential reasoning, or a form of *Naturphilosophie*.[12] He inferred that electrical forces were mediated through an "ether," which was an unidentified medium whose properties were unknown.[13] The relevance of Ørsted's deduction that an electrical ether existed becomes clear when considering that prior to that time, electricity was viewed by scientists as analogous to a fluid—a force they could comprehend. Following Ørsted's work, familiar physical analogies were no longer sufficient. The Franklinian view of electricity as a fluid was understandable by lay society, but Ørsted's discoveries demonstrated that electricity had properties the fluid analogy could no longer explain, that there were forces involved that transcended simple comprehension. While historians view Ørsted's work as a transformative point in the practical development of electrical technology, his theoretical discoveries indicated that electrical force, as an invisible entity, would be too complicated for the public to grasp.[14]

A year after Ørsted demonstrated the link between magnetism and electrical current, Michael Faraday was beginning to conduct experiments at the Royal Institution in London.[15] Ørsted had first proven the relationship between magnetism and electricity; Faraday demonstrated how magnets could be used to create electrical current. Beginning by thrusting a bar magnet through a wire coil, Faraday demonstrated that an electrical current was created in the wire, which could be recorded on a galvanometer connected to the wire coil. He also showed how electrical current could be induced into a wire wrapped around an iron ring that was also wrapped with a separate wire through which current flowed. In 1831, as Faraday became more interested in electromagnetism, he developed the homopolar generator, also known as the Faraday Disk. Faraday's device utilized a magnetic disk that rotated within a magnetic field, or between two stationary magnets, to produce an electrical current.[16] The Faraday Disk was the precursor to the modern dynamo in that it converted mechanical energy manifested in the spinning disk to electrical energy, which is the basis for all nonphotovoltaic modern electrical generation.

As with Ørsted before him, as critical as Faraday's invention was from the standpoint of generating electricity, his theoretical work is important in understanding how lay society came to comprehend—or more fittingly, not comprehend—what electricity was. Whereas Ørsted theorized that electricity was a force that acted upon an electrical ether, Faraday's field theory eliminated the theory of ether altogether. In 1846, Faraday sought to "dismiss the aether, but not the vibrations," which theoretically made the conceptualization of electricity even more abstract than Ørsted had made it twenty-five years earlier.[17] Faraday's theoretical model completely departed from those that had come before. In the linguistic metaphors of

Franklin's fluid model and Ørsted's concept of the ether, the exigencies involved a mechanical model in which electrical force flowed through or rode upon a medium.[18] The broad concept of mechanical models of force, in which the ether served as an effluvium on which various phenomena such as light could be transmitted, had originated with the theories of René Descartes. For Isaac Newton, the ether was the active and ubiquitous agent on which the force of gravity could act.[19] As the first nonmechanical model of electricity, Faraday's conceptualization illustrated just how different electricity was in comparison to other forces that society experienced. As the technological knowledge of electricity evolved with Faraday, it became harder for a nontechnical world to intuitively define electricity as a knowable entity. Considering that theorists from Descartes to Newton to Ørsted all struggled to comprehend phenomena without a mechanical analogy, it stood to reason that ordinary people would also have difficulty understanding how to define electricity. Due in part to these issues of definition, electrical forces came to be culturally constructed as either ethereal, mysterious, or remaining grounded in Franklin's logical fluid model that manifested itself in terms of the "flow" of electricity. While Faraday's technical work led directly to the practical application of electrical power on display in Philadelphia in 1876, the complexity of his theoretical discoveries illustrates why electricity became disassociated from nature by the general public—it was just too complicated for most to comprehend.

At around the same time Faraday began working in England, Joseph Henry, a professor in Albany, New York, had also begun experimenting with electromagnetism.[20] As early as 1831, Henry described an electric motor that he had developed:

> I have lately succeeded in producing motion in a little machine by a power, which, I believe, has never before been applied in mechanics by magnetic attraction and repulsion. Not much importance, however, is attached to the invention, since the article, in its present state, can only be considered a philosophical toy; although, in the progress of discovery and invention, it is not impossible that the same principle, or some modification of it on

a more extended scale, may hereafter be applied to some useful purpose. But without reference to its practical utility, and only viewed as a hew effect produced by one of the most mysterious agents of nature, you will not, perhaps, think the following account of it unworthy of a place in the Journal of Science.[21]

Not only did Henry downplay the significance of the practical application of his invention, he also referred to electricity as "one of the most mysterious agents of nature," showing that even among persons of science, the phenomenon was difficult to conceptualize. As significant as Henry's development of the first electric motor was, his early work also demonstrated the viability of transmitting power over long distances.[22] Henry's improvements on the crude electromagnetics that existed in the early 1800s made possible the first electromagnetic telegraph prior to Morse's.[23] Henry's telegraph was in use in 1831, in his classroom at the Albany Academy, where he had strung a wire of nearly a mile in length over which he successfully transmitted a signal.[24] Among historians, Henry is widely credited with the invention of the telegraph, although Morse received full credit for the invention in American popular culture.[25]

As groundbreaking as Henry was in the development of modern electrical technologies, one of his most significant insights regarding electricity and energy in general has not been mentioned in the historical record. Henry knew that energy production was at its core a series of tradeoffs, stating in 1875:

> All attempts to substitute electricity or magnetism for coal power must be unsuccessful, since these powers tend to an equilibrium from which they can only be disturbed by the application of another power which is the equivalent of that which they can subsequently exhibit. Science does not indicate in the slightest degree, the possibility of the discovery of a new primary power comparable with that of combustion as exhibited by the burning of coal. Whatever unknown powers may exist in nature capable of doing work, must be in a state of neutralization, otherwise they would manifest themselves spontaneously; and from this state of neutralization or equilibrium, they can be released only by the action of an extraneous power of equivalent energy; and therefore

we do not hesitate to say that all declarations of the discovery of a new power which is to supersede the use of coal as a motive power have their origin in ignorance, deception, and frequently in both.[26]

While Henry did not foresee the development of nuclear, water, solar, or viable wind power in his assessment, he did understand that electrical energy generated by coal should not be considered a "new" power, going as far as saying that it was ignorant to declare that it was. His assertion that unknown powers can be released only by the actions of an extraneous power also anticipated modern debates over alternative energy sources such as ethanol, gasification, and other motive energy sources that can often require as much power to produce as they yield.[27] Henry was the first to recognize the concept of energy abstraction by implying that the substitution of electricity for other motive forces was merely a deception.

As a scientist and teacher, Henry did not pursue the commercialization of his work, nor did he recognize the potential of his discoveries at the time. He felt that steam engines were more practical than any electric motor could be, and he did not pursue the development of the telegraph outside of his classroom.[28] Henry, like Faraday and Franklin before him, was not particularly interested in the practical application or commercialization of his work, claiming, "the only reward I ever expected was the consciousness of advancing science."[29] What Henry did accomplish, however, was to show that electrical power, both as a motive force and as a method of communication, could be transmitted via wire, and his work served as an important link between the theoretical and the practical application of electrical power.

While Henry's work did sow the seeds of electrical technologies that began to modify the spatial dynamics of motive power and communications, it was Morse who commercialized the first technology to transmit electricity over long distances via wire. Whether or not Morse conceived of the electromagnetic telegraph independently has been the subject of a long-running debate, but he was issued the first US patent for the device and the Morse code system in 1840.[30] In addition to Henry's work, several European inventors, as well as William Fothergill Cooke and Charles Wheatstone, predated Morse's patent for telegraphic technology.[31] Regardless of who was the first to conceive of the telegraph, Morse's single-wire system was inexpensive to implement, included a workable language in the Morse code system, and most importantly, incorporated relays that allowed electrical signals to remain strong over long distances. Morse had most likely learned of the idea for relays from none other than Henry, who in 1837 invented a crude relay, which he referred to as an "intensity magnet."[32] Although Morse later denied that Henry had anything to do with his invention of the telegraph, Morse's notes from March 1839 indicate that Henry shared his thoughts about boosting signals to work at distances of one hundred miles or more.[33]

Jean Antoine Nollet established that wire could transport electricity over a distance in 1746. In a well-known experiment, Nollet lined up two hundred Carthusian monks in their monastery, had them hold hands, and had the first and last monk grasp the ends of a long wire connected to a charged Leyden jar. Upon the liberation of the static electricity from the jar, the monks simultaneously jumped, proving that current travelled not only via wire but also through any suitable conductor.[34] However, the issue of resistance or loss of current as it passes through a conductor remained problematic for transmission over distances.[35] Morse solved the problem of resistance by incorporating relays that boosted electric telegraph signals by the application of a fresh charge of electrical power at specific distances along the transmission wire.[36] Overlooked in significance by many, the electromagnetic relay is one of the most critical components of Morse's system, as it allowed electrical communications over unlimited distances.[37] The combination of components—Morse's electromagnetic telegraph system, batteries to power the system, wires to connect points, and relays to allow electrical signals to travel unlimited distances—represented the first instance of electrical power altering the spatial dynamics of existing functions, in this case, communications. Morse's work did not transport electrical energy that could be employed for illumination or motive power, but it was significant because it represented the first application in which wire abstracted process.

When Morse constructed a forty-mile-long network between Washington, DC, and Baltimore in 1844, and invited Miss Annie Elsworth to send the first public message from the chamber of the US Supreme Court to Morse's colleague Alfred Vail in Baltimore on May 24, she sent the now famous message, "What hath God wrought."[38] Elsworth's choice of words not only delighted Morse but represented the association of the ethereal and spiritual with the force of electrical transmission by wire—an association that continued as long-distance transmission of electricity advanced. On May 31, the *Baltimore Sun* commented: "Prof. Morse's Telegraph has already, during the first week of its operations, been proved to be of the greatest public importance. Time and space has been completely annihilated."[39]

By the time of the Centennial Exhibition in Philadelphia, Morse's telegraph had matured to the point that Edison was introducing improvements on Morse's design by increasing the speed of sending and receiving via a duplexing system. Because Edison's first encounter with electrical technologies came when he was working as a telegraph operator in Mount Clemens, Michigan, in 1863, it was his exposure to Morse's system that spurred his interest in the possibilities of the transmission of electricity by wire.[40] The Morse telegraph was the first practical application of electrical energy transmitted via wire in the United States, and it was followed in 1876 by Bell's telephone, introduced at the Philadelphia fair. Despite the enthusiasm for the phone shown by Dom Pedro III of Brazil and others who witnessed Bell's demonstration at Machinery Hall, the initial social reaction to the telephone was tepid at best. Communication historian Richard John acknowledges that "with the exception of trade journals and chronicles of notable inventions, the telephone rarely received more than a passing mention in the press."[41] The telephone did become widely adopted by American society, but not for another half a century after its introduction. The telegraph remained a business communications tool not suited for common social networking or home use. While the telephone did not represent a technology that filled an existing household need at the time, the possibility of an incandescent light did. The potential for a practical technology that could replace open flames for illumination indoors was realized by a number of entrepreneurs and anticipated by the press.[42]

Lighting systems in place at the fair in 1876, such as Wallace and Farmer's and Gramme's, were of the arc light variety, and were limited in their application. The carbon rods that produced light within an arc system required frequent replacement, were prone to buzzing and flickering, and produced an intensely bright white light. Arc lights were commercially successful for outdoor lighting, as some localities found them less expensive for street lighting than gaslights, but for indoor use they equated to an unpleasant fire hazard. The potential practicality of arc lighting outdoors as an alternative to natural gas seemed to dominate newly established electric trade publications, with some even speculating that the role of gas would one day be relegated to supplying only "motive power for dynamo-electric machines."[43] Despite the disadvantages of arc lighting, new installations in cities large and small were popular spectacles that drew significant crowds.[44] From all indications, the majority of the public was enthusiastic over the coming of the electric light, even in its earliest iterations.

Charles F. Brush of Cleveland, who was present at the Centennial Exhibition, developed the first widely adopted arc lighting system in the United States. Two of the initial problems with these systems were determining how to power multiple lights with a single generator and developing a regulator or feed mechanism that could advance the burning carbon electrodes. Brush developed a complete system that included regulating mechanisms and lights wired in series to matched dynamos allowing for multiple lamps to be powered by one central system.[45] Brush's first significant installation of his lighting system was in April 1879 in Cleveland, utilizing a generation system that he financed. Shortly after, the first investor-owned electrical power network, that of the California Electric Light Company, went online in September of the same year using Brush dynamos.[46] The first publicly owned electrical utility system in the United States went online in Wabash, Indiana, on March 31, 1880, with ten thousand spectators present for the inaugural lighting. The Wabash *Plain Dealer* wrote of the occasion:

> The people, almost with bated breath, stood overwhelmed with awe, as if in the presence of the supernatural. The strange, weird light, exceeded in power only by the sun yet mild as moonlight, rendered the Court House square as light as midday. While we contemplated the new wonder in modern science, we could but think how our electricians had got it on Ben Franklin. He brought down the lightning from the heavens on a kite-string and bottled it, just to show, presumably, how smart he was. Brush and Edison take a steam engine, belt it to a huge electro-magnetic machine, manufacture lightning and use it to light cities and hamlets, thus benefiting mankind and blessing posterity.[47]

Similar to the comments expressing exuberance for electrical technology following the Centennial Exhibition in 1876, the lighting of Wabash was described in terms of the supernatural and the ethereal, and also equated with the tenets of Progressivism in terms of "benefiting mankind and blessing prosperity." Although the Brush generator was close to the lights, only as far away as the basement of the very courthouse it lighted, the enthusiasm for the system overshadowed any consideration of the electrical current's origin. A witness describing the generator observed:

> From these lamps the spectator will notice two ordinary telegraph-size copper wires leading down over the roof and down the west side of the building to the basement, where stands the Brush Dynamo Electric Machine that generates the current of electricity that flows through the wires to the carbons, between which it flashes with the brilliancy of lightning. The leaping of this current from one carbon pencil to the other produces the light, and the space thus made brilliant is termed the voltaic arc. This dynamo machine occupies a space of four feet in length and two in width and will last for years. It is practically indestructible, all its wheels

revolving in the air. It requires no chemicals, and generates the most powerful electricity.[48]

In this discourse, the dynamo machine is revered much like the lighting itself, "indestructible" and requiring "no chemicals." The description of the current as "leaping" indicates that the writer was struggling to define electricity's properties, searching for analogies as Franklin and Ørsted had a century earlier. The Brush dynamo was powered by an old threshing machine steam engine on the courthouse lawn with a pile of coal adjacent to it.[49] Although smoke was problematic in big cities, there is no mention in the historical record of a smoke nuisance from Brush's Wabash installation. In the various newspaper accounts of the occasion in Wabash, no mention of wires, engines, or coal is found, suggesting that no thought was given to the origins of the power that ran the lamps.

The Brush arc lighting system gained in popularity during the 1870s and by 1881 the company reported installations worldwide, including some 800 lights in various industrial facilities; over 1,200 in woolen and cotton mills; 425 in stores, churches, and hotels; and 1,200 lights in various municipal applications, not to mention installations in mines, railroad depots, and other establishments.[50] In Manhattan by 1880, Brush lights had been illuminating Broadway some two years before Edison's first generating plant went online. Although there were other lighting companies proposing their own systems, Brush dominated the headlines because he had established installations in place, at the same time Edison had been reporting publicly that he was closing in on the invention of a viable incandescent lighting system. In December 1880, the *Cleveland Plain Dealer* ran an article reporting on the rivalry to light New York with the subheadline, "What Brush Has Done and Edison Proposes to Do."[51] Mentioning Edison's unfulfilled proposition was not surprising. A man who welcomed press coverage, Edison had claimed two years earlier that he had solved the problem of the electric light by inventing a practical incandescent bulb. Under the headline "Edison's Newest Marvel," the *New York Sun* printed an article in September 1878 in which Edison claimed that it would be possible to send "cheap light, heat, and power by electricity."[52]

In promising more than he could deliver, at least at that moment in time, Edison built enthusiasm for electric power, which was rapidly catching the public's attention. Edison had taken a different tack than his rivals—he was looking to replace the gaslight system, and his model for electric light and power distribution was that of the existing gas companies. In the same *Sun* article, Edison is quoted as saying that his new system would make the use of gas for illumination a "thing of the past." The fact that Edison was thinking about replacing existing gas systems is significant because it indicates that his vision was firmly rooted in widespread distribution. Although Brush systems were installed in Cleveland, Wabash, New York, and San Francisco, by the nature of the technology only a very few arc lights could be powered at once. The San Francisco power plant, of the California Electric Light Company, powered only twenty-one arc lamps that were rented to customers for $10 per week.[53] While the Brush systems did rely on central generators, they were limited in the number of power-hungry arc lights they could supply. Edison was proposing a replacement for gas illumination not just outdoors or for large installations, but for interior household use as well. An opportunist, Edison aimed to replace gas lighting in every home, and in the process eliminate fire-based lighting technology for the first time in history. Claiming that he could light "the entire lower part of New York City using one 500 horse power engine," Edison's sights were set on larger distribution systems that would be inexpensive and accessible. This was a vision encompassing more than just lighting. Edison claimed that "the same wire that brings light to you will bring you power and heat."[54] Further promoting his system, Edison described water- or steam-driven dynamos that could transport power from a distance. Although he would not have an operational generation and distribution system for another four years, Edison started to identify electricity as a panacea by claiming that his electrical system could "run an elevator, a sewing machine, or any other mechanical contrivance that requires a motor, and by means of the heat you may cook your food."⁵⁵ With his assertions, Edison began to position electricity as a modern and miraculous power source, an idea that would remain embedded in the zeitgeist of American culture.

In the context of the late nineteenth century, Edison's claims positioned electricity as the antithesis of the existing technology of smoke and fire. During the same month in which Edison announced that he had solved the problem of the incandescent light and could potentially deliver heat via wire, newspapers were reporting on the dangers associated with lamps, candles, stoves, and fires. Articles with headlines such as "Candles Are Dangerous," which reported on a fire from a candle in a New Orleans home, or "Fatal Coal-Oil Fire," which told of a young woman's death in Washington resulting from a can of spilled oil on a heating stove, were typical.⁵⁶ Insurance records documented hundreds of fires resulting from gas lamps, defective flues, candles, and kerosene lanterns; at least one fire insurance company in Boston began to work with a lantern manufacturer to design a safer means of illumination.⁵⁷ Against this backdrop of hazards, "Edison the Magician" was self-assuredly announcing that he could eliminate the gas lamp and even the coal or wood stove.⁵⁸ Although there were those in the scientific community who were skeptical of Edison's claims, ardent believers were mesmerized by the prospects of electricity in general, and rested assured that the "Wizard of Menlo Park" was hard at work. It was just a matter of time before the incandescent bulb and the magic of electricity would enter their homes and liberate society from the grips of fire.⁵⁹

Confidence in Edison's ability to deliver a suitable replacement for the gaslight was such that just the announcement of the technology sent gas stocks plunging, and on January 27, 1880, his incandescent bulb came to fruition with the issuance of US Patent 223,898, for an electric lamp.⁶⁰ Edison was not the first to invent an incandescent lamp—most notably, Joseph Swan in England had developed a lightbulb that he announced on December 18, 1878, but because he delayed filing, his British patent was not issued until November 1880.⁶¹ Even if Swan had patented and marketed his bulb earlier, Edison's stature, reputation, and press coverage gave him momentum—and the funding needed to make his system operational.

From 1880 into 1882 Edison developed the components necessary to build his ultimate objective, an electrical generation system in the form of a central station supply model. A central station could distribute electrical power to the public, in contrast to dedicated generating plants or isolated stations used only by their owners. Edison's central station supply model was the early precursor to the more modern electrical grid, and he decided to build his first station at 257 Pearl Street, in New York City's financial district. This decision was the result of two considerations. From a marketing standpoint, the location was near Wall Street and potential investors, and from a technical standpoint, the nature of Edison's system was that it produced direct current (DC) power, which did not lend itself to long-distance transmission.

While the newly formed Edison Illuminating Company worked to build and wire the Pearl Street Station in 1881, J. P. Morgan, one of Edison's largest investors, insisted that his residence receive electric lighting before anyone else. Edison was not a fan of private or dedicated generating plants, yet he committed to installing incandescent lights in the Morgan residence. Morgan's excitement was evident, as he requested that the company proceed "with all possible dispatch" in a letter he wrote to Calvin Goddard, the secretary of the Edison Company.[62] By June 1882, workers had installed a coal-fired generator in a basement under the stables at Morgan's mansion at 219 Madison Avenue.[63] While Morgan was delighted with the lighting and the operation of the system, his neighbors were not. In December, James M. Brown, who lived behind Morgan, complained of smoke and noise from the generator, and in response Morgan pressured the Edison Company to fix the problem.[64] A similar incident occurred with the Vanderbilt family and an Edison installation at its home in Manhattan. The newly remodeled mansion was equipped with Edison's lights, and in a large picture gallery with a silk wall covering interwoven with metal threads, a small fire occurred during the initial lighting of the lamps. Mrs. Alva Vanderbilt, who became hysterical, demanded to know where the electricity was being generated, and when she was told that there was a boiler and generator in the basement she said that she would not occupy the house until it was removed. "She would not live over a boiler," according to Edison biographer Frank Dyer.[65] These stories illustrate that when electrical generation is close to the point of use—when the space between generation and consumption is condensed—cognizance of fire, smoke, and steam is omnipresent. In both of these cases, electricity was not truly invisible, as the detritus from its generation was experiential and obvious. This soon changed.

Edison's station plant on Pearl Street began to generate power on September 4, 1882, and was widely recognized as the first central power plant in the United States.[66] By 1883 it powered 8,573 lamps in Manhattan.[67] Under Edison's original two-wire system, distribution of electricity farther than a mile from the generating plant was not feasible due to the nature of direct current and the size requirements of the copper mains (the thick wires that transported power), which were too heavy for overhead use and cost-prohibitive. Edison eventually adopted a three-wire system that allowed for smaller cables and longer transmission of direct current power, yet the practical range was limited to less than ten miles unless the voltage of the current was substantially increased.[68] While by contemporary standards the Pearl Street plant was limited in the distance it could transmit power, it nevertheless established a new model of central station distribution and began to shift the way energy was delivered and conceptualized in the United States. Physically, the Edison system inserted more space between power generation and consumption, and allowed for the replacement of the gas line or steam pipe with wire. Instead of a burning gas jet, lantern, or candle in the home, Edison and Pearl Street represented a significant step in moving flame and fuel to the central station where it could be forgotten. Conceptually, Edison's system and his rhetoric began to position electricity as progressive and gas and coal as regressive. When Edison announced that "the same wire that brings light to you will bring you power and heat," he began to alter the way the public viewed energy.[69] With his announcement, Edison not only foretold of a future where the spatial dynamics of energy delivery would be radically different, he eliminated the salience and burden of energy consumption. Departing from the paradigm of the past, Edison spoke of a world where energy was effortlessly and passively *brought* to the consumer via wire, eliminating the need to go *get* coal or kerosene, or to actively light a flame. Suddenly, the narrative changed from one of fuel to one of consumables. In the past wood, coal, or oil was the deliverable; now it was light, heat, and power itself.

The overall development of deployed electric power came in several stages. Brush's arc lighting systems, and those of others, represented the first wave; second came Edison's system of distribution and incandescent lighting. The third wave followed in the form of alternating current technology. While the Pearl Street Station was expanding the number of lights and locations it supplied power to in 1883, two articles, one in the English *Engineering* magazine and the other in *Scientific American*, described a new system of power generation and distribution, "the Gaulard-Gibbs Secondary Generator."[70] As a condition of the "practical employment of electricity," *Scientific American* cited the necessity of the ability for "the distribution over great distances, not of electrical current, but of electrical energy, which is an entirely different matter."[71] The article compared Edison's direct current power with the Gaulard-Gibbs system, which was a practical means of transmitting alternating current power. A direct electrical current flows in one direction and cannot be easily stepped up or down in voltage. What this means for transmission is that direct current power is less versatile in the way it can be transported via wire and consumed. Alternating current electricity changes directions, and its voltage may be stepped up or down utilizing a transformer, which is the system that Gaulard and Gibbs developed in 1883.[72] The utility patent issued to Gaulard and Gibbs in 1886 describes the technological improvement over existing electrical distribution systems as the ability "to convey a useful quantity of electric energy to a much greater distance than has heretofore been practicable, while the cost of the necessary plant for electric lighting and other analogous purposes, especially that of the main electrical conductors, is very materially diminished."[73] Also of interest in the patent is the name of the assignee, the American industrialist George Westinghouse, who owned the prosperous Westinghouse Air Brake Company in Pittsburgh. He became interested in the Gaulard-Gibbs distribution system and bought the US patent rights, along with the patent rights of several other electricity-related technologies including Joseph Swan's lighting technology and generator patents. As the direct current arc lighting market was competitive, and Edison's direct current incandescent light system had technological and market momentum on its side, Westinghouse saw an opportunity in alternating current power—especially in its ability to be transmitted over greater distances.

While several European companies had been looking at alternating current power systems that utilized Swan incandescent lamps, no alternating current systems were in place in the United States and the Edison system had virtually no competition. On March 20, 1886, Westinghouse engineers completed work on a small pilot project in Great Barrington, Massachusetts, and turned on the lights. It was the beginning of a system of electrification that would change the shape of US society.[74] The ensuing "war of the currents," which pitted Westinghouse's alternating current system against Edison's and other direct current systems in the marketplace, determined not only the shape of the power grid but also how modern America would define electrical energy.[75] In the context of the spatial dynamics between power generation and power consumption, the lighting of the first store in Great Barrington was the beginning of a paradigm shift in the way electrical power was delivered and subsequently perceived.

In the ten years between 1876 and 1886, electrical power in the United States moved from an experimental to an applied technology. While the electrical technologies on display in 1876 at the Centennial Exhibition in Philadelphia represented systems that began to abstract energy by transporting invisible forces by wire, they were largely novelty technologies, not yet adopted by society at large. Mysterious and hard to define, the properties of electricity eluded all who encountered it, from those such as Franklin and Ørsted to commentators in the 1800s who continued to conceptualize the technology in deistic and Progressive terms. With the enthusiasm built by early promoters of electricity, most prominently Edison, many wrote of it as a miraculous force, especially when set against a backdrop of fire and smoke.[76] As both the theoretical and practical work on electricity progressed after 1876, the process of delivering usable energy continued to alter perceptions and in turn obfuscate electricity's ultimate source of generation. As central supply stations began to increase the distance between the generation and the consumption of energy more than ever before, the disassociation between electricity, coal, and fire increased as well. Culturally, electricity began to be seen as an energy panacea equated with American progress.

CHAPTER 6
ENERGY, UTOPIA, AND THE AMERICAN MIND

> Between the dynamo in the gallery of machines and the engine-house outside, the break in continuity amounted to abysmal fracture of a historian's objects. No more relation could he discover between the steam and the electric current than between the Cross and the cathedral.
>
> Henry Adams

In his classic 1964 work, *The Machine in the Garden*, historian Leo Marx observes that mechanization gradually became part of the pastoral image of America as the public imagination was captivated by technologies that represented progress. Marx's hypothesis came from his analysis of literature and art, and he pointed out that even Emerson wrote that "Machinery and Transcendentalism agree well."[1] Four years later, in *Wilderness and the American Mind*, Roderick Nash took a similar approach regarding the social construction of wilderness with his well-known line claiming that the wilderness was made by "the literary gentlemen wielding a pen, not the pioneer with his axe."[2] Similar to Marx's characterizations of mechanization and Nash's ideas about wilderness, modern attitudes toward energy in general, and electricity in particular, were formed in the last two decades of the nineteenth century through the influence of public events, in the press, and in literature. Just as ideas about race, gender, and disabilities are social constructions, so are perceptions of energy.

Attitudes about electricity began to form with Americans' first exposure to electrical technologies. As a nebulous, invisible energy source that society found hard to define, the abstraction of electricity began with the invention of the telegraph in 1848, accelerated after the world's fair of 1876, and shifted again between 1882 and 1900. In the last two decades of the century, the perception of electricity changed from a mysterious entity to a utopian energy source. The public began to see the systems and technologies on display at the fair in 1876 as solutions to problems associated with energy derived from old-world fire and smoke. The attitudinal shift that occurred was due both to advances in electrification technology itself and to cultural influences derived from events, press coverage, and literature. Technologically, the adoption of alternating current, the electrical technologies on display at the Chicago World's Fair in 1893, and the much-anticipated opening of the Niagara Falls generating station in 1895 deeply affected public attitudes. Culturally, press coverage of these events, along with positive portrayals in the popular genre of utopian literature, depicted electricity as a panacea—an unlimited, progressive power source with no deleterious consequences. The first stage of energy abstraction occurred with electricity's midcentury American debut; the next stage emerged after 1882, as technological advancement drove more practical applications and allowed for greater space between generation and consumption.

As successful as Thomas Edison's Pearl Street Station was at delivering on his promise to illuminate parts of midtown in 1882, the inherent flaw in his direct current system remained. To provide power to all of Manhattan, or any other large city, a plant would have to be placed every few miles to transmit power to the businesses and residences that wished to consume it. Edison's original plan included thirty-six independent coal-burning power stations to supply power to central New York City alone.[3] Recognizing this weakness in Edison's direct current system, George Westinghouse aggressively pursued alternating current after his engineers implemented the technology in 1886. Westinghouse's strategy was not without risk: the original alternating current power was single phase, which had some technical limitations, and there was no electric motor yet perfected that would run on alternating current power.

Two years prior to Westinghouse's successful pilot project in Great Barrington, the Serbian inventor Nikola Tesla had moved to New York to take a job with Edison. Tesla had worked for an Edison affiliate in France and was recommended to Edison by a colleague, Charles Batchelor. As a young engineer in Budapest, Tesla had experimented with and developed concepts for a polyphase alternating current system, including an electric motor and an efficient generating system, but Edison, who had invested money and pride in his direct current systems, would not consider any of Tesla's ideas regarding the superiority of alternating current. Assigned to work on improving Edison's "Jumbo" direct current generators, Tesla lasted six months in his contentious employment under Edison. After working for a year as a ditch digger in New York, Tesla met Charles F. Peck and Alfred S. Brown, two investors who were interested in electricity as a business, who assisted him in setting up the Tesla Electric Company in 1887.[4] Peck and Brown were interested not just in electricity but in energy in a broader sense, including geothermal energy from the ocean.[5] While working on several alternative energy projects for Peck and Brown, Tesla perfected his "Electro Magnetic Motor," which he patented in May 1888 along with three other inventions for the transmission of alternating current electrical power.[6] In July, Tesla, Peck, and Brown sold their patent portfolio to Westinghouse, who was ready to move forward with larger distribution systems and the deployment of alternating current power.

The Westinghouse/Tesla partnership was key in the development of the modern electrical grid, but not without an initial struggle with Edison, who tried to discredit alternating current technology. The "battle of the currents" between Edison and Tesla has been written about at length in both histories of technology and studies of public relations. An overview of the controversy begins with Edison's market share in the deployment of direct-current-generating installations across the country being threatened by the Westinghouse alternating current system, and Edison fighting back with a negative public relations campaign that attempted to paint alternating current technology as dangerous. Edison's smear campaign went as far as lobbying the state of New York to accept electrocution as a method of capital punishment, using the Westinghouse system. A colleague of Edison's, Harold P. Brown, convinced state authorities in New York in 1889 that it would be a humane method of death. Both Brown and Edison testified in Cayuga County Court that the current was so lethal that death would be instant.[7] Edison had already publicly electrocuted horses, dogs, and an elephant to demonstrate the negatives of alternating current, yet the Westinghouse interests argued correctly that direct current power at the same voltage and amperage would be lethal as well. In public, the battle between Edison and Westinghouse played out in the press, with Edison arguing in the *North American Review* that his system was absolutely safe and that the alternating current system of Westinghouse was deadly.[8] Edison did have his own credibility on his side, and his direct current system as deployed *was* fundamentally safer than the Westinghouse alternating current system. The Edison system in Manhattan was referred to at the time as a "low tension" system, which meant it was low voltage, transmitted at less than two hundred volts—shocking, for sure, but not instantly lethal. Alternating current systems, however, including those of Brush, Westinghouse, and all other arc light businesses, were normally transmitted at over a thousand volts, which, while not necessarily lethal, would cause visible effects such as burning, arcing, and other injurious manifestations. While Edison and Brown spun the issue to be over the merits of alternating current versus direct current, it was more a debate over high-voltage versus low-voltage as well as variances in amperage. In addition to Edison's credibility as an influential factor in the debate, there had been a number of accidental electrocutions of line workers in the 1880s, and since all alternating current systems used overhead wires, these were often public spectacles where electricity suddenly became nonabstract and quite salient.[9] Between Edison's credibility and a well-orchestrated public relations campaign by Westinghouse's competitors, the New York prison system decided that alternating current power, deemed the "executioner's current" by Brown, would be an instant and humane method of capital punishment.[10] On August 6, 1890, the first execution in an electric chair, of convicted murderer William Kemmler, was anything but instant or humane. The *New York Times* headline, "Far Worse than Hanging," needed no explanation.[11]

While this episode has been framed by historians as an illustration of Edison's mercurial personality as well as a story of the struggle between emerging standards, it is perhaps more relevant in the study of how the public came to view electricity. Americans' acceptance of electricity was not without setbacks. The "wire panic" in New York City occurred when several horrendous accidental electrocutions of workers led to a brief period of technological pessimism in 1889.[12] Regardless of these incidents, however, the public came to accept electricity as a beneficial and abstract technology—horrendous death was possible with any technology, from railroads to industrial accidents—but the sensational accounts of accidental or irresponsible death did not faze the public's perception. Less than three months after Kemmler's execution, the *Electrical Engineer* reported that the Westinghouse Electric and Manufacturing Company would have the best year it had ever had, with new installations of alternating current systems in more than ten cities.[13] In 1893, twenty-seven million Americans traveled to Chicago to marvel at the lights of the Columbian Exposition world's fair—powered by alternating current technology.[14] The Kemmler episode and the well-publicized accidental electrocutions in New York demonstrated that the public's confidence in electricity was not easily swayed. Despite Edison's best efforts to discredit Westinghouse, the inherent disadvantages of Edison's system, primarily in its inability to transmit power efficiently at distances more than about a mile, worked in Westinghouse's favor.

Between 1887 and 1890, the Westinghouse Company continued to compete with Edison's direct current systems, and despite financial difficulties due to rapid expansion, alternating current systems slowly began to take market share. In 1887, Westinghouse systems supplied power for 134,000 incandescent lamps; by 1890 the number had grown to half a million, with 300 central stations producing power.[15] By the end of 1890 alternating current systems were gaining preference and the Edison Company was losing money. In his December 1890 essay published by the *North American Review*, Westinghouse succinctly explained the benefits of the alternating current system and the deficiencies of direct current, pointing out that alternating current was preferred by a five-to-one margin.[16] Edison General Electric stock plummeted. Edison's investors, including J. P. Morgan and Anthony Drexel, had already merged several of Edison's companies into the Edison General Electric company in 1889, and although the company was still pushing direct current, management was demanding that Edison work on a competing alternating current system. By 1892 Morgan merged Edison General Electric with another, more profitable competitor, Thomson-Houston, changed the name of the new company to General Electric, and forced Edison out as a principal. Edison owned stock in General Electric, but was no longer active in the company. Commenting to the *New York Times*, Edison said, "I cannot waste my time over electric lighting matters, for they are old. I ceased to worry about these things ten years ago."[17] While Edison's departure from General Electric was a significant step in the demise of direct current power, two other events that transpired afterward marked the final blow, one being Westinghouse's winning bid to provide lighting for the Columbian Exposition in Chicago in 1893 and the other being the decision by the developers of the Niagara Falls generating station to use Westinghouse/Tesla technology.

If the Corliss engine was the heart of the world's fair in Philadelphia in 1876, then the electric light was at the center of the fair in Chicago in 1893. If Philadelphia represented the beginning of a process of technological abstraction, where power generation and consumption were first separated by emerging technical systems, then Chicago represented a new level of technological fantasy, in which a white city and white lights eclipsed the realities of burning coal. Historian Robert Rydell's assessment, that the "effort by America's leaders to define social reality reached a new level of sophistication with the Chicago World's Columbian Exposition of 1893," is telling, because electrical power was featured and was squarely at the center of the definition of what an ideal society should be.[18] Organized to celebrate the four hundredth anniversary of Christopher Columbus's landfall in the New World, the Chicago planners were determined to outdo anything that had come before. Covering an area of 633 acres, the fairground site was at Jackson Park, where architects Daniel H. Burnham and John W. Root designed the fair's neoclassical buildings. Painted in white, the compound became commonly known as the White City. The fair's opening on May 1, 1893, was initiated by President Grover Cleveland, who, after a speech, pressed a button that started an Allis steam engine, much like the opening of the fair of 1876 when President Grant had started the giant Corliss steam engine. Setting the Columbian Exposition apart from Philadelphia, the *Chicago Tribune* said of the fair's opening:

> This dramatic ceremony will bear little resemblance to the touching of the button by President Grant at the opening of the Centennial Exhibition at Philadelphia. In the first place, though it was popularly believed at the time that by this act he started up the Corliss engine, it is now reported that he only rang a signal bell, and that the engineer opened the throttle by hand. In the second place, the Corliss engine furnished all the power and operated all of the equipment in the Centennial Exhibition, while the Allis engine, though much larger than the Corliss, does not furnish more than one-twentieth of the power required in the World's Columbian Exposition.[19]

Here, the reporter made an effort to outdo the past, and in the context of technology. The article borders on ridiculing the nonautomated world of the past. Not only was Grant's engine starting flawed, but once the Corliss did start, it paled in comparison to the new two-thousand-horsepower Allis-Corliss engine in Chicago—and even though it was bigger, the Allis still was not big enough to supply the power for the fair, its function being only to run a pump for the water fountains as well as two dynamos that could power twenty thousand lightbulbs near the fountains.[20] Philadelphia here is framed as old steam, and Chicago as new and electric. Rydell describes the fairs in terms of "symbolic universes," and as such, the symbolism at Chicago was one of progress, power, and electricity intertwined.[21]

While the *Tribune* reporter highlighted the Allis-Corliss engine at the center of the building known as the Palace of Mechanical Arts (commonly referred to as the machinery building), the article leaves out the real heft driving the fair. The main power plant, removed from the primary exhibit hall, covered a space one hundred feet wide and one thousand feet long and housed seventy-seven engines, nine of which were devoted to turning generators that produced electricity to run up to 120,000 Westinghouse-provided lights.[22] The biggest engine was an E. P. Allis quadruple expansion condensing steam engine capable of generating three thousand horsepower. Another twenty-three engines drove smaller generators to power outdoor arc lighting. In 1893, journalist John Patrick Barrett wrote of "dynamos of all conceivable kinds . . . which were divided into two classes, those producing direct or continuous current and those generating alternating currents. Late advances made in electrical science permit the use of either kind for the same purposes, but for the utilization of electrical energy at any conceivable distance from the source of power, the alternating current system possesses advantages of flexibility that make its use imperative."[23] Barrett's comments, reinforced by others, not only highlight the promise of transporting energy over great distances but also speak to the flexibility and advantages of electricity—and deservedly so. Just seventeen years after Philadelphia, the Chicago fair featured tens of thousands of lights, numerous electric motors, a complete fair telephone exchange, fire and police alarm technologies, grand visual displays, the Edison "Tower of Light," and thirty-eight thousand colored arc lights shining over rising and falling jets of water in the center pavilion.[24] The Westinghouse display, along with a number of novelty lights, featured a likeness of Christopher Columbus outlined with small incandescent bulbs.[25] The White City had arrived, and it was progressive, modern, and above all else, electric.

The official attendance at the fair in 1893 was approximately 40 percent of the US population at the time.[26] The White City was a harbinger of modernity for rural and urban visitors alike, with electricity at its center. Journalist Teresa Dean wrote in her diary that she heard a man say, "I tell everyone in my town that they must come to the Fair. And if they can't get the money to come any other way, they better knock a man down gently and take his money, and then after they return from the Exposition, go to work and pay back in installments the man they've robbed."²⁷ Writer Hamlin Garland famously wrote to his parents, "Sell the cook stove if necessary and come, you must see this fair."²⁸ The *Chicago Record* reported that a woman from Texas named Mrs. Lucille Rodney walked from Galveston on railroad ties, a distance of thirteen hundred miles, to get to Chicago.²⁹ The *Chicago Daily Tribune* reported on the opening of the Electricity Building at the fair as "overpowering in its magnificence, rivaling nature in the variety of her wealth and color."³⁰ Poet Daniel Oscar Loy, struck by the building's luminescence, wrote:

> In the Electric Building
> I tarried for an hour,
> Learning all there is to learn
> About electric power.
> I heard Thomas Edison
> Speaking of his latest light,
> Which is as bright as the sun
> Making day out of night.³¹

The *Rock Island Daily Argus* reported that the fair was "the climax in electricity's upward march through the nineteenth century,"³² while the *Bismarck Weekly Tribune* declared the 1,250,000 candlepower of lighting as "so complete and extensive" that it was "well worth the journey to see."³³ Ralph Pope, secretary of the American Institute of Electrical Engineers at the time, mentioned that faith in electricity was all-consuming, that "people have got an idea that electricity can do anything."³⁴ Dean noted that "in the Electricity Building, which was brilliantly lighted . . . we went there and stood looking at the electric picture of Columbus."³⁵ A correspondent with the *Omaha Daily Bee* reported on housewives who saw electric ovens "without the semblance of a spark or fire," and "little wires that run to irons for laundry purposes."³⁶ Just three years after

Edison's attempt to discredit alternating current technology with the Kemmler execution and various scares over wires and electrocution, in Chicago and around the country, the public had come to embrace electric power.

While the public face of technology had changed between 1876 and 1893, behind the scenes the fundamentals were familiar. Far removed from the dynamos and the power plant was an iron structure that housed over forty immense boilers that supplied steam for all of the engines, including those that ran the dynamos in the power plant. The steam, born of immense boilers and transported underground, was ultimately created by oil-burning fire, produced in the fire-room nine feet below the boilers.[37] The oil that produced the heat, which made the steam that drove the engines and dynamos, ultimately resulting in displays of clean electricity, was pumped to the fire room from storage tanks a half mile away.[38] The fuel oil itself was likely Ohio oil, refined in Whiting, Indiana, where John Rockefeller's Standard Oil Company had recently completed a refinery to serve the Chicago market.[39] One contemporary report boasted that "there is no smoke, dust, or dirt as there would be if coal were burned."[40] Reinforcing the imperative that the White City—a symbolic universe—must be free of smoke, a sophisticated system was installed that included an "inspector of smoke," who was stationed in a cabin near the main oil valves yet in view of the fire-room chimneys. In the case of a chimney emitting smoke, the inspector could push a button—one for each set of boilers—that would vibrate a gong near the specific boiler, which then would alert the fireman to attend to his fires, regulating the oil flow to reduce the smoke. Considering that those observing the exhibits saw only clean white lights and quiet electric motors, the operation behind the scenes was abstraction realized, a technological sleight of hand that obscured energy generation from consumption, thus reinforcing inconsequentiality. Even at its point of creation, electricity is shrouded as a secondary source of energy. Electricity's journey begins once it has been removed from the forces that ultimately create it, as dynamos must always be turned by other energy forms. It is here that the direct line back to the primary energy source quickly starts to blur. For the public, the electrical technologies represented modernity and progress. Electricity came from dynamos—not from the oil fields of western Ohio, and not from the fires beneath the boiler room.

Cutting through the celebratory rhetoric and the complexity of the hidden infrastructure, there were a few dissenting voices. Henry Adams, historian and great-grandson to John Adams, had been fascinated with electricity and wrote regarding the Chicago fair in 1893 and the Paris exposition of 1900 that electricity was "but an ingenious channel for conveying somewhere the heat latent in a few tons of coal hidden in a dirty engine house kept carefully out of sight."[41] Whether or not Adams's ambivalence toward technology was the antimodernist rant of a "displaced patrician" has been debated, and although he stood mostly alone in his skepticism over electricity, he did keenly observe the problem.[42] Adams's "dirty engine house kept carefully out of sight" came to define electrification in the United States in the forthcoming century. The ideal city, as embodied in the White City of Chicago, moved the engine house away from the lights, buried the fires in the ground, and posted smoke spotters to make sure any mischief exiting the chimneys was quickly reined in. Beneath it all was the paradigm of steam, but soon this too changed.

Winning the contract to supply the lighting and power for the world's fair in Chicago vindicated Tesla's technical foresight and Westinghouse's vision for the future of electrical transmission. But while the alternating current systems in place at Chicago might have captured the imagination of the American public, signaling that the age of electricity had arrived, the promise of unlimited electric power from Niagara Falls left a larger legacy.[43] The promise of Niagara Falls as a source of electrical power generation had been talked about since the early 1880s. Although no practical plans for large-scale generating stations were in the works until 1886, some were already speculating that Niagara could supply enough electricity for all of North America. In an article published by the *Chicago Daily Tribune* in June 1881, Sir William Thomson, now Lord Kelvin, conjectured that "Niagara [was] the natural and proper chief motor for the whole of the North American Continent; and it now seems quite within the bounds of possibility that people who are now living may witness the application of this chief motor to the indicated uses."[44] Lord Kelvin visualized the use of batteries, which were not perfected at the time of his writing, that could store power from Niagara and possibly be shipped to other cities via trains to supply power. He was a staunch direct current advocate, and although he did envision transporting electricity over long distances, his idea was to physically move the power to where it was needed. In addition to the power that Niagara could supply, Lord Kelvin mused of atmospheres: "Smokeless and clean, uncontaminated with the products of combustion; with flowers and fruit flourishing in town gardens; with our rooms, and especially our public rooms and places of assembly, freed from the heat which gas occasions; and with nature and art manifest in their true colors by night as well as by day."[45]

In effect, Lord Kelvin was describing an electrical utopia powered completely by the natural force of Niagara Falls. The significance that he attached to the grand cataract continued to grow as journalists and others reluctantly recognized that the world's coal fields might be exhaustible, and therefore alternatives such as power from Niagara should be approached with a "practical interest."[46] In short, Niagara became idealized and hydropower became the foundation on which a newly imagined electric future could be built. Within the broader context of late nineteenth-century Progressive America, the promise of Niagara represented another step in human mastery over nature, just as Jacob Bigelow had envisioned in 1829. Ever since Francis Bacon expressed the idea that "the empire of man over things is founded on the arts and sciences alone," Western civilization had found advancement wrapped in the systematic exploitation of natural forces, and in the late 1800s the timing was right for Niagara. Optimistic about an imminent era defined by technological progress, Americans saw firsthand—in Philadelphia and then in Chicago—the possibilities of an electric future. Although the giant Corliss in 1876 and banks of dynamos in 1893 represented mechanical perfection, it was only because the fire and steam that drove them were tucked away in a "dirty engine house," part of an inconvenient truth that was easily ignored.

In 1886, it was Thomas Evershed, an engineer who had worked on the Erie Canal, who first outlined a large-scale plan for harnessing the power of the falls. His plan was to bore deep vertical shafts at a point in the upper Niagara River west of the city of Niagara Falls that would channel water downward into a deep tunnel. The tunnel would run for over two miles, beneath the city, and through various wheels with shafts extending upward through the ground the power of the falls could be captured for mechanical power to be supplied to hundreds of mills. Local promoters, enthusiastic over Evershed's plan, acquired the requisite property but ran short on capital and were forced to sell their holdings.[47] Three years later, a group of investors led by New York banker Edward Dean Adams and backed by J. P. Morgan formed the Niagara Falls Power Company. Whereas Evershed's original plan was to exploit the power of the falls to run hundreds of mills with mechanical energy, Adams's group concluded that the power of the falls was best captured at a central station to generate electricity. The original plan called for transmission of the power to Buffalo, which was largely a speculative technology at the time because long-distance transmission of electrical power had not yet been perfected. In 1890, workers began digging tunnels, and in 1891, Niagara Falls Power sought plans for the best system of hydroelectric power generation at the falls. Although the planning phase of electrical generation at the falls occurred in the midst of the Edison-Westinghouse debates over direct and alternating current, the company selected Tesla's alternating current. As construction of the power station commenced during the Columbian Exposition in Chicago, excitement as to the potential for Niagara power grew. Tesla himself had predicted that the electricity generated at Niagara could provide power around the world, with the potential for running streetcars in London and streetlights in Paris. In his enthusiasm, Tesla claimed that "humanity will be like an ant heap stirred up with a stick. See the excitement coming!"[48] At the official opening ceremony of the Niagara hydroelectric plant, Tesla, as had so many others before him, spoke of electricity in religious terms, adding that man's subjugation of nature would save humanity:

> We have many a monument of past ages; we have the palaces and pyramids, the temples of the Greek and the cathedrals of Christendom. In them is exemplified the power of men, the greatness of nations, the love of art and religious devotion. But the monument at Niagara has something of its own, more in accord with our present thoughts and tendencies. It is a monument worthy of our scientific age, a true monument of enlightenment and of peace. It signifies the subjugation of natural forces to the service of man, the discontinuance of barbarous methods, the relieving of millions from want and suffering.[49]

Tesla was not alone in his excitement over the potential of Niagara. In the media, Niagara took on a larger-than-life role in the future of the country, with stories of electrical force so great that it was "impossible to conceive what would be possible by its application . . . turning out invisible force to give life to the factories and railroads."[50] Other stories, picking up on earlier visions of coal-free power, explained that "the line of the roof of the [Niagara] power station is unbroken by chimneys. This is because the building is heated throughout by electricity."[51] Speculation that Niagara could displace coal, that "its daily force was equal to the latent power of all the coal-mines in the world each day," was not unheard of.[52] Although the electricity generated at Niagara was at first sent via wire only to Buffalo, twenty-six miles away, there was growing speculation that harnessing the falls to supply more distant locations was a real possibility. Even a press report of "an atom" of Niagara power that had been transmitted around the world via telegraph line was cause for national coverage and great excitement.[53] Tesla and others believed that Niagara's power would make Buffalo the "greatest city in the world," a phrase that Westinghouse eventually adopted in its advertising.[54]

In its own way, the establishment of power generation at Niagara had a substantial impact on the American public's attitudes toward electricity. At the world's fair in Philadelphia in 1876 the public discourse portrayed electricity in terms of amazement and mystery. By the early 1880s Edison's commercialization of the lightbulb and lighting systems led to a portrayal of the inventor in terms of practicality and progress. Chicago's White City of 1893 impressed upon visitors that the ideal city ran on clean electricity, with endless possibilities. Now, with Niagara online, a cultural custom was being established in the press that endless coal-free electrical energy could be transmitted anywhere. At Niagara it seemed that humankind had mastered nature and tamed the mysterious force of electricity for the good of society, a panacea realized. In an age of smoke, Niagara represented the possibilities for alternative energy, an extinguished flame, a realization of Edison's earlier promise that power for light, heat, and cooking would all be delivered into one's home by wire. Between the White City of Chicago and the promise of Niagara, the cultural construction of electricity as a utopian and progressive force for the future now accelerated. American English began to incorporate expressions that could not have existed prior to the introduction of electricity; words such as "human dynamo," "electrifying," and "shocked" began to appear in a number of publications and advertisements.[55] In literature, works that incorporated themes of energy and electricity were nothing new, but began to shift as the promise of invisible energy appeared on society's horizon.

Electricity as a mysterious force had made its literary debut in London in 1818 with the publication of *Frankenstein, or, The Modern Prometheus* by Mary Shelley. At the time of Shelley's writing, galvanism, or animal electricity, was a popular topic in London, inspired by the work of Luigi Galvani and his experiments with frogs' legs and electric shock. In *Frankenstein*, Shelley uses an electric shock as the vital force of life, an idea that captured the imagination of both English and American readers in the early nineteenth century.[56] By 1851, Herman Melville incorporated the enigmatic force of energy into American literature in *Moby Dick*. For Melville's antagonist Captain Ahab, it is the electrically charged spark of lightning that represents omnipotent force. Ahab is so captivated by the white light of lightning that his first mate Starbuck must pull him out of his trance and back to the harpoon and the hunt—yet it is electricity that will guide him and "light the way to the white whale."[57] In Mark Twain's satire *A Connecticut Yankee in King Arthur's Court* (1889), the American Hank Morgan travels backward in time after a blow to the head and attempts to modernize sixth-century England by using his knowledge of the future. During his exploits, Morgan performs feats of magic through the use of technology. One of Morgan's first acts is to build an electric plant in Merlin the magician's cave, with which he electrifies fences for protection, runs wires for dynamite charges, and lights up castles to the amazement of King Arthur's subjects.[58] Twain, a person who was fascinated with electricity—to the extent that he allowed the passage of a current through his body when visiting Tesla's laboratory—also includes an offhand explanation of proper wiring and grounding in his novel.[59]

While the works of Shelley, Melville, and Twain took advantage of the mysterious nature of electricity as a force, American utopian and dystopian novels in the late 1800s illustrate the confidence assigned to electricity as a progressive energy source. Literature scholar Jean Pfaelzer has argued that the "nineteenth-century utopian novel . . . can hardly be understood as a serious prediction of historical process."[60] While this may be true in the realm of economic development and politics, several works do foretell of advances based on emerging technological developments. In a time of great social change, the utopian novel was a literary expression of the author's anxieties, and smoke, steam, and energy played important roles in the most popular works of the late nineteenth century. Kenneth Roemer observes that "coal, soot, and other odor-producing fuels" were commonly replaced by electricity in utopian works between 1888 and 1900, and along with aluminum and high-speed rail, electricity was the most mentioned technology.[61]

An analysis of this popular genre not only validates Roemer's observations but reveals utopian settings that mirror cultural perceptions derived from the White City and Niagara, both of which were characterized as clean electrified spaces. An example of this portrayal is found in the most popular work of the period, Edward Bellamy's 1887 publication *Looking Backward: 2000–1887*.[62] As Bellamy's protagonist Julian West finds when he visits the year 2000, smoke and fire are gone. His host from the future, Mrs. Leete, explains, "Electricity of course, takes the place of all fires," and thus she not only removes flames and smoke in the forthcoming world, she positions electricity as a sole power source.[63] Through Mrs. Leete, Bellamy reflects an early manifestation of energy abstraction. Writing at a time when all electricity was derived from fire and steam, Bellamy reinforced the disconnection between coal and current by promoting a position that dismissed the link between power generation and consumable energy. In the utopian world of Mrs. Leete, electrification represented the new, the clean, and the future—and it was antithetical to fire. As part of Bellamy's utopian ideal, technological advancements were assumed to be safe and part of a more humane, orderly, and civilized future. Bellamy's lesser-known sequel, *Equality*, published in 1897, continues with similar themes, although electricity plays an even bigger role. In *Equality*, West witnesses electric plows and motors connected by a system of flexible cables, electric cars for travel, and a possible precursor to the internet in electrically connected "electroscope" networks.[64] As in *Looking Backward*, Bellamy's sequel reinforces the role of electricity as a replacement for whale oil and as a successor to steam; although *Equality* was written during the time of Niagara's development, there is no mention of the source of the electricity.[65]

In 1890, Populist political leader Ignatius Donnelly published *Caesar's Column: A Story of the Twentieth Century*, which describes electricity as a force that has been conquered, as well as a force on which "the happiness of millions depends."[66] Extending Shelley's depiction of electricity as life-giving, Donnelly wrote of a future in which the "slow process of agriculture would be largely discarded, and the food of man would be created out of the chemical elements of which it was composed, [then] transfused by electricity and magnetism."[67] In Donnelly's future world, the technology of electricity is far more advanced, dynamos are replaced by the "magnetism of the planet itself," there are electric magazines, and electric air transports that consist of "huge, cigar-shaped balloons, unattached to the earth."[68] By removing dynamos from the future, Donnelly not only increased the distance between generation and consumption, he removed generation completely, making the derivation of electrical energy completely inconsequential. Clearly, in this scenario electricity becomes the transformative technology, yet since Donnelly's novel is a dystopian rant against capitalism, only the ruling elite truly enjoy the spoils of technology. As with Bellamy's work, Donnelly's novel is a cautionary tale against the excesses of capitalism, and along with equating electricity with technological advances, both works portray electricity as an egalitarian social force, in contrast to Gilded Age coal-based capitalism.

Donnelly and Bellamy were not alone in associating electricity with a cleaner, better future. William Dean Howells, one of the most influential and widely circulated authors of the period, also saw electricity as part of an improved future.[69] In Howells's utopian *A Traveler from Altruria*, published in 1894, "the capitals are clean," partly due to "electrical expresses that transport the artist, the scientist, and the literary man."[70] While Howells was cleaning up cities with electrical expresses, in *The Human Drift*, author King Camp Gillette envisioned a modern world powered by electricity derived solely from hydropower. Clearly motivated by the excitement over the coming of Niagara, Gillette's utopian city of Metropolis is completely powered by hydropower-driven dynamos. Located "about ten miles east of Niagara and Buffalo," Metropolis includes not only manufacturing centers but also luxury apartments that are "heated and cooled by automatic mechanisms, lighted by electricity, and electrically connected with the whole outside world."[71] Gillette's work serves as another example of coal, smoke, and steam as the antipode of the utopian space.

In feminist utopian works of the time, electricity plays a commanding technological role as well. Mary Bradley Lane's publication *Mizora: A World of Women* (1881), includes carriages propelled by compressed air and electricity, and since Mizora is a haven in the center of the earth, the "dreamy daylight" is produced by electricity.[72] As in Donnelly's work, the

Mizorians in Lane's novel rely on electricity to produce their food, utilizing electricity, carbonic acid gas, and hothouses to grow fruits and vegetables. In Mizora electricity sustains life, yet the precise source of it is unclear. Anna Bowman Dodd's *The Republic of the Future, Or, Socialism a Reality*, from 1887, uses electricity to send food great distances through "culinary conduits" and to run all of the machinery in the home.[73] As a visitor to the year 2050, Dodd's protagonist, Wolfgang, writes to his friend Christina, who lives in 1887: "I had noticed almost immediately on my arrival that throughout the city, not a chimney was to be seen. It naturally followed that, there being no chimneys, there was also no smoke, which therefore made this already sufficiently clear atmosphere as pure as the air on a mountaintop."[74]

Throughout these works, the utopian worlds are egalitarian, communalistic, and above all, electric. While these portrayals reflect a style of technological utopianism that positions coal and steam as technologies of a dystopian world, they are also forced to remove dynamos and further abstract energy and electrical generation. Unlike any of the other works discussed here, William Dean Howells came close to recognizing this issue when his protagonist, Mr. Homos from the utopian island of Altruria, explains: "It was long before we came to realize that in the depths of our steamships were those who fed the fires with their lives, and that our mines from which we dug our wealth were the graves of those who had died to the free light and air, without finding the rest of death. We did not see that the machines for saving labor were monsters that devoured women and children, and wasted men at the bidding of the power which no man must touch."[75] As with many of these other works, Howells does not address the source of Altruria's electric power, but the passage above does start to connect consequence to abstracted energy. The depths of the steamships, the mines far away, and the machines not seen in Howells's world directly equate to Adam's "dirty engine house" in Chicago and the growing space between power generation and consumption that are about to follow. While Howells begins to identify the issue, he falls into the same trap as the other works—their utopias need electricity, but there is no utopian method to provide the power they need. As a result, the future worlds deal with the provision of electricity through the means of power magically derived from the earth or atmosphere, or generation is not dealt with at all, suggesting to the reader that dynamos are unnecessary and will long be a thing of the past by the year 2000.

Beginning in the mid-1880s, the development and gradual adoption of alternating current technology allowed for increasingly larger distribution systems that placed more physical space between the generation and the consumption of electricity. Fully realized at the Chicago Columbian Exposition in 1893, this and other advances in electrical engineering created an imagined White City, void of steam and smoke, which reflected the possibilities of technological utopianism. As Chicago captivated both the public and the press, the promise of unlimited clean power from Niagara Falls contributed to a public sentiment that positioned electricity apart from coal, further abstracting the dynamo from usable electric power. In the press, speculation that electricity could be stored and would eliminate fire and that the ideal city could be realized continued to contribute to the idea that electricity was an energy panacea. By the end of the nineteenth century, societal views of electricity had undergone a shift. What was once a magical novelty was becoming a force that represented mankind's mastery over nature and was a social solution to smoke- and fire-based drudgery. In end-of-the-century American utopian novels, electricity as the featured technology solved problems of food production, transportation, and coal-based capitalism. While the forward-looking literature of the day influenced society's view of electricity as a savior, it also promoted the myths that grew out of the White City and the anticipation of Niagara, of unlimited electrical energy with no coal and no consequences. The idea that electricity was an exceptional energy source disassociated from smoke and fire remained part of American culture as the modern electrical grid began to take shape.

CHAPTER 7
TURBINES, COAL, AND CONVENIENCE

> There will be no further need of digging dirty coal, for cheap and clean electricity will light and warm the world and furnish motive power.
>
> Walter J. Ballard

Driving by 1111 W. Cermak Road in the Pilsen neighborhood of Chicago today is an uneventful experience. Situated between the South Branch of the Chicago River and a single-track railroad siding is an abandoned power plant once known as the Fisk Street Station. With locked chain-link gates and a smokestack void of emissions, the plant is now quiet. The last load of coal delivered from Wyoming's Powder River coal region by rail arrived in midsummer of 2012, more than a century after the plant first went into service. At the time of its closing, the *Chicago Tribune* quoted the director of the Environmental Law and Policy Center in Chicago as saying that the closing "marks a turning point from Chicago's reliance on two highly polluting coal plants that use fuel from out of state to a cleaner energy future that's less polluting and uses more Illinois wind and other clean resources."[1] Celebrated in terms of environmental inconsequentiality when it opened in 1903, reporters praised the coal-fired plant in much the same way as they did Niagara Falls few years earlier. The plant was deemed "smokeless" and heralded as "One of the World's Seven Wonders" that would "diminish smoke throughout the city" due to its 205-foot smokestacks.[2] The Fisk Street Station was the brainchild of Samuel Insull, a protégé of Thomas Edison. While Edison's first plant in Manhattan twenty-one years earlier was significant as the first attempt at central station generation, Fisk Street had a far greater effect on the future of energy in the country. Fisk Street marked the beginning of an electrical generation and transmission paradigm that made coal-derived energy invisible, established coal as the nation's prime mover, and contributed to the belief that electricity was a clean and modern technology.

The physical structure of the modern electrical grid began to emerge in the early decades of the twentieth century. Along with the establishment of technological systems that would remain in place, social views of energy consequentiality and consumption had already begun to form and were equally enduring. Having passed through stages of energy abstraction that went from the mysterious to the utopian, electricity was widely adopted in American society, eventually becoming the sine qua non of progress and modernity. As the twentieth century began, a radical disruption occurred within previous models of power generation and distribution by the development of steam turbines and the deployment of large regional transmission systems made possible by alternating current. This emerging paradigm became the model for energy distribution for the next one hundred years and further separated power generation from power consumption geographically.

Simultaneous to the technical turn, a cultural shift was under way. As forces of consumerism and Progressivism took hold in an expanding American middle class, an ever-increasing faith in technology along with the rise of advertising positioned electrification as a gateway to modernity. With the demand for electricity growing and the economies of scale made possible by steam turbine power plants such as Fisk Street, the cost of electricity decreased and energy consumption rose dramatically. By 1930, over 80 percent of households in the nation were electrified, the conscious disassociation of coal from electricity accelerated, and as electricity became inextricably tied to American consumer culture, unlimited power consumption was encouraged.[3] While both marketers and intellectuals passed along celebratory cultural messages informing the public that the Age of Electricity had arrived and the Age of Coal had passed, the nation consumed more coal than ever before.

As the twentieth century began, electrification in the United States was in a state of flux. Since the start-up of Edison's

Pearl Street Station in 1882, the methods and technology behind electrical generation and transmission had evolved asynchronously, which led to systemic discontinuities. Characterized by a mix of existing direct current systems and emerging alternating current systems, electric power distribution was a nested structure of subsystems, characterized by what historian Thomas Hughes refers to as "reverse salients."[4] Hughes defines these as components in a system that do not "march along harmoniously" with other components.[5] If the system is to proceed, the reverse salient requires correction or resolution. In the closing years of the nineteenth century, expansion of electrification was hampered due to its basis on an Edisonian direct current paradigm, characterized by small municipal or private power plants providing electrical power to confined areas. The main focus for Edison, Westinghouse, and others was to sell equipment, lightbulbs, appliances, power generation systems, and franchises. Within these small systems, electrical current was often sold on a per-lamp basis—a carryover from the captive arc-lighting systems of Brush and others.[6] The shift away from direct current began in 1893 with Tesla's innovations, the adoption of alternating current technology at Niagara, and the subsequent beginnings of long-distance transmission of electrical power. While these technical advances were significant, the method of power generation went largely unchanged, and electricity remained an energy source primarily derived from the burning of coal.

Between 1894 and 1912, a radical disruption in technology altered the way electricity was generated and distributed in the United States. In the process, power generation became farther removed from power consumption, the electrification infrastructure became less visible, and coal became established as the primary fuel source for the generation of electricity for a century to follow. While names such as Thomas Edison, George Westinghouse, and Nikola Tesla loom large in the history of technology, the long-term impact of Samuel Insull had a greater effect on how Americans perceived and consumed electricity. Under the direction of Insull, Chicago Edison built a model of power generation, distribution, and marketing that the entire country eventually adopted, displacing existing systems that had been in operation since the 1880s. This process of change began at the Fisk Street Station.

Prior to the late 1890s, electricity in the country was produced by generators driven by coal-fired reciprocating steam engines, an engine design that had been fundamentally unchanged since its invention by Matthew Boulton and James Watt a century earlier. Although there are significant differences in reciprocating steam engines, at the core are pistons and valves, and the familiar up-and-down motion that is converted to a circular motion as in a locomotive's drive wheels. In the early days of power generation, the rotary motion of the steam engine was connected to a generator with a drive belt that turned an armature to create electrical current. The first electrical generation stations were of this design, including Edison's Pearl Street Station in Manhattan. From a technological standpoint, this pairing resulted from the fact that reciprocating steam engines were established prior to the invention of the generator, and the two contrivances became adapted to create a power generation system. At best, reciprocating designs were 20 percent efficient, meaning that only 20 percent of coal's chemical energy was converted into reciprocating motion.[7] The lack of efficiency meant that more coal had to be burned to produce a megawatt of electricity, and since the variable cost of plant operation is mostly in fuel, the cost of electricity was high.

In 1884, Charles Parsons of England perfected a steam turbine generator or "turbogenerator" that altered the calculus of coal-to-electricity efficiency. Compared to a reciprocating engine, a steam turbine is smaller and lighter per unit of horsepower, and rotates at a higher speed, gaining efficiencies of up to 80 percent.[8] The operation of a steam turbine is straightforward: injected under pressure, steam flows onto enclosed rotor blades, causing them to spin (the principle is broadly similar to that of a pinwheel, where moving air creates rotary motion). Because there is no conversion from reciprocal to rotary motion, and because more of the potential thermal energy can be utilized, turbines gain both efficiency and speed. The turbine design eliminated a separate engine and drive belt system entirely, as steam-driven turbine blades were integral to the generator's shafting. All modern power plants use steam turbines—coal, nuclear, or natural gas, and all are methods of heating water to produce steam to drive turbines. One of the first successful installations of a steam turbine was that of a small Parson-type seventy-five-kilowatt generator in England at Newcastle in 1890, and the first large-scale generation plant went online in 1899 in Eberfield, Germany, with a capacity of 1 megawatt, followed by 2.5-megawatt units in Frankfurt in 1901.[9] In 1895, New York Edison's West 39th Street plant installed the first steam turbine in the United States.[10] While the first American-built units were small, with capabilities to generate about 500 kilowatts of power, by 1900 Westinghouse had manufactured and installed a medium-capacity 1,500-kilowatt unit at the Hartford Electric Company in Connecticut.[11] (One megawatt [MW] is equal to one thousand kilowatts [kW].) These two early US installations represented an experimental stage of turbine technology; it was not until 1903 that Insull deployed turbines on a large scale at the Fisk Street Station.

Insull did not invent the steam turbine, nor was he the first to utilize it, but he was the first to build high-capacity regional central stations. Insull also consolidated small neighborhood stations, which were artifacts of the Edisonian direct current model and leveraged the economies of scale that resulted from larger generation plants. When Insull became the president of Chicago Edison in 1892, the technological momentum for the wider adoption of electricity was well under way, and few were more experienced in the burgeoning electric power industry at the time. Starting his career as Thomas Edison's secretary in the 1880s, Insull was present for the start-up of the Pearl Street operation and he remained in New York until J. P. Morgan consolidated Edison's business and transformed it into General Electric. In 1892 Insull sought new opportunities and interviewed with the board of directors of the fledgling Chicago Edison Company for the position of president. Despite the impression of an electric utopia at the Chicago Columbian Exposition in 1893, outside of the fairgrounds the electrical infrastructure of Chicago was patchwork, with more than forty-five electric companies operating independently.[12] As one of the forty-five operations, Chicago Edison was a small player in a disjointed infrastructure whose territory covered fifty-six square blocks in the downtown loop district.[13] A shrewd businessman who was able to take advantage of the economic downturn in 1893, Insull bought a number of competitors in the Chicago market, and by 1898 Chicago Edison had a virtual monopoly on electrical generation in the nation's second-largest city.[14] Insull was not a crusader with a desire to provide a cleaner, safer source of power to the masses; he was cut from the mold of Gilded Age capitalists, poised to exploit a new technology as profitably as possible. At the core of Insull's strategy was scale: larger generation plants that could produce more power at a lower cost. While this yielded more profit for Chicago Edison, it also allowed the company to sell electricity to the consumer at a lower cost.

Insull's focus on maximizing profitability in electrical generation was obvious, and along with many other power plant operators, he believed that coal generation was the most profitable method, even before the implementation of steam turbines. Although Niagara had demonstrated the feasibility of hydropower, the electric interests had always favored coal as the most cost-effective way to generate power. At the National Electric Light Association convention in 1898, Insull listened as Mr. W. M. Walbank presented a paper on the cost of producing electricity by hydropower at the Lachine Rapids installation in Montreal.[15] After Walbank explained various aspects, including capital costs to build the plant, generation capacity, and cost of water rights, he concluded, "From the foregoing, the writer trusts that he has shown that where reliable water power can be obtained within reasonable distance from power centres it can be made to produce cheap electric current, to say nothing of the great advantages the city must derive therefrom, not only commercially, but viewed from a sanitary standpoint as well, as the use of electric power thus generated is the best smoke consumer yet invented."[16] The debate that ensued after Walbank's presentation centered on the relative cost and merit of electrical generation with a hydropower plant versus a steam and coal plant. American utility interests in the room espoused the lower cost of steam; standard steam plants at the time cost much less to build than hydropower plants, and the cost of the coal fuel was offset by the lower capital costs of construction. The issue of smoke—Walbank's mention of the sanitary standpoint—was not addressed in the debate. Comments by Insull and other plant owners focused purely on cost to produce a kilowatt-hour of electricity.

With a keen focus on profits, Insull was naturally interested in the efficiency possible in steam turbines, and during a 1901 European vacation, he first saw the large German installations.[17] Intrigued by the efficiency and potential profitability of the units, Insull approached General Electric (GE) about the possibility of supplying him with a five-megawatt steam turbine for a plant Chicago Edison was planning to build. Although GE had encouraged him to take on a smaller, one-megawatt unit, Insull persisted and guaranteed to take a portion of the risk if the unproven design failed.[18] In the fall of 1903, the five-megawatt turbine went online at Fisk Street, and within a year and a half, Insull scrapped the five-megawatt units for turbines of thirty-five megawatts. By 1906 the total output of the station was 156 megawatts.[19] Insull realized that lowering the cost of producing electricity not only resulted in greater profits but also gave Chicago Edison the ability to offer electricity at lower rates than competitors and allowed him to market his energy as inexpensive. The adoption of steam turbines for generation led to a major shift in how electricity was produced, and Insull's move set the trend. Within a year after the construction of the Fisk Street plant, General Electric and Westinghouse had manufactured and sold steam turbines across the country that represented a total generation capacity of 540 megawatts.[20] The age of the coal-fired steam turbine for the generation of electricity had arrived.

Just four years after Fisk Street went online in 1903, the cost of coal as a percentage of total operating costs of large central generation stations had dropped by 3 percent, a trend that continued as turbines increased in size.[21] In addition to the cost of coal itself, the economies of scale achieved from the steam turbines came from a number of factors, including reduced capital costs relative to generation capacity and plant efficiency.[22] Before the advent of steam turbines, reciprocating-piston-type steam engines achieved a thermal efficiency of 3 to 5 percent. Simply stated, this means that only 3–5 percent of the heat energy produced is utilized for work, while the remaining 95–97 percent is wasted. Early steam turbines improved thermal efficiency by a factor of three to five times by achieving efficiencies of 15 percent.[23] For those considering relative costs of alternative fuels such as petroleum, natural gas, and ethanol, which were rarely used at the time, nothing came close to coal for low-cost energy generation.[24] A US Department of Agriculture report from 1908 found that "it was possible to buy eight times as much energy in the form of cheap coal" when compared to most other fuels.[25] Although hydropower garnered considerable excitement in the press and among engineers who were attracted to a potential fuel cost of zero, energy executives were well aware of the high cost of hydropower plant construction, maintenance, and the interest on debt to finance their construction.[26] In addition, those in the power industry feared interruption by natural forces such as drought or floods and laws that discouraged the development of hydropower. Legal obstacles such as gaining permission for transmission right-of-ways, public domain rights, and other legislative hurdles were burdens not inherent in unregulated steam plant construction.[27] With coal plants already being less expensive to build and more profitable to operate, the consideration of mine-mouth plants—steam turbine plants built directly proximal to coal mines—further added to the promise of coal for future profits by eliminating the freight costs of coal.[28]

As coal cost and supply became critical to the operation of profitable electrical generation, Insull invested in coal to ensure a stable supply. Francis Peabody, an aspiring coal magnate in the early years of the twentieth century, needed considerable capital to grow his company. Insull had capital and needed coal. The two men struck an agreement in 1913 for Peabody to supply Chicago Edison with all the coal that it needed for the foreseeable future at cost, plus a small profit.[29] With the contract in hand, Peabody bought additional mines to supply the coal that Insull required. Whereas the power companies may have had the biggest incentive to use coal due to its profitability, capitalists such as Insull and Peabody were not the only parties responsible for locking in coal as the nation's ultimate source of electricity. Mine owners, especially Peabody, and labor made sure that the coal supply was steady and inexpensive. After Peabody won his first contract to supply Chicago Edison with coal in 1913 for half a million tons per year from mines in southern Illinois, he negotiated with John L. Lewis of the Illinois mine workers' union.[30] Peabody agreed to support safety laws for the mines and in exchange Lewis agreed that all contracts between the mines and the union would expire on April 30 of each year—just prior to the summer months, when the demand for coal and electricity was at its lowest.[31] From the late 1890s until World War I, the average price of bituminous coal rose more slowly than the wholesale price index and remained lower and more stable than crude petroleum and anthracite coal.[32]

The profitability realized due to the construction of large steam turbine plants and inexpensive coal was occurring not only in Chicago. In 1906, reciprocating steam engines began to be replaced by five-kilowatt steam turbines at the Twin City Rapid Transit Company plant in Minneapolis, building on the success of installations in Chicago and New York.[33] In the emerging power grid in the United States, larger turbine plants served more distant regions, further separating electrical generation from consumers.[34] For example, in 1913 a five-megawatt steam turbine plant in Missouri Valley, Iowa, displaced three smaller, unprofitable regional plants. A similar coal-burning plant built near Galena, Illinois, served customers in a two-hundred-square-mile area. Missouri's Empire District Electric Company began servicing a scattered population of over 150,000 people and 165 miles of interurban railway via a central station with over one hundred miles of high-voltage transmission lines.[35] In all of these instances, economies of scale realized by efficient steam turbines, an inexpensive coal supply, and capital costs spread over an increasing customer base kept the cost of electrical production down and helped to embed a coal-based model that became widely adopted during the early twentieth century. While independent operators duplicated Insull's model of larger plants, Insull himself expanded his holdings—by 1907 he owned twenty additional utility companies and exploited the scalability of steam turbine generators and alternating current transmission to control the entire electric supply in Chicago. In 1911 he established the Public Service Company of Northern Illinois and controlled thirty-nine more electric companies; in 1912 he formed Middle West Utilities to acquire more electric companies throughout the Midwest.[36] By the second decade of the twentieth century the steam turbine had completely changed the calculus of electrical generation in the country. Insull established the model for large regional systems serving consumers increasingly distant from the plants that supplied them with power, and the nascent power grid evolved with coal and steam at the core.

As the total electrical generating capacity in the United States increased from 2,987 megawatts in 1902 to 10,980 by 1912, the majority of the increase came from coal-fired turbine plants, firmly indicating the acceptance of the technology. The adoption of the steam turbine, along with the confidence in the future of hydropower, came at the expense of research into alternate sources of energy. As John Adolphus Etzler in the early 1800s and John Ericsson in the mid-nineteenth century discovered, power companies had no incentive to pursue anything other than power generation by large coal or water turbine installations. Along with the socially driven momentum of electrification, the steam turbine had taken on a momentum of its own, as research and development concentrated on engineering bigger and more efficient turbines. Insull, for example, scrapped the original turbines installed at Fisk Street in favor of larger turbines when "the progress of the art was such that practically the same boiler room arrangements" were able to operate turbines with a much higher capacity.[37] Even though there is evidence of interest in pursuing wind-driven electricity generation, neither General Electric nor Westinghouse—the two major generator manufacturers in the country—had any reason to pursue it.

Ironically, it was in the West, atop some of the nation's largest coal reserves, where one of the most serious early pursuits of wind power occurred. In April 1913, Frank Bosler, a rancher outside of Laramie, wrote to General Electric's manufacturing division to inquire about harnessing the wind to generate electricity for a regional power system he planned to build. Responding to Bosler, the company stated that "they had not developed any particular apparatus for wind generation," and were positive that wind would not work for electricity generation due to the inconsistent nature of wind speeds.[38] Bosler responded that the wind in Laramie was constant, and he pursued his idea by exploring the possibility of rigging together windmills with belt-driven dynamos and using batteries to store power. A letter from the Electric Storage Battery Company of Philadelphia to Bosler echoes the position of General Electric, that wind was an "unreliable energy source."[39] Bosler continued to experiment with wind power until 1916, after which time he began to pursue both hydropower and coal-fired plants. With the adoption of coal-fired steam turbine technology on a broad scale, the development of and interest in alternative energy sources such as wind were largely dismissed for nearly a century.

By 1912, every state in the country with the exception of Delaware and Utah had operable central station coal-burning steam turbine generation plants, and five states—Illinois, New York, California, Pennsylvania, and Massachusetts—garnered the majority of their electricity from large turbine plants.[40] As turbine sizes increased, the number of coal-fired steam generation plants decreased as per-plant capacity increased—this did not mean that less coal was consumed; it simply meant that more coal was consumed in fewer plants. Because of their capacity, coupled with alternating current, turbines allowed for the building of bigger generation plants, which in turn could serve more consumers in a wider geographic area. Older, inefficient plants were converted to substations, which did not generate power but stepped it down to lower voltages appropriate for domestic use.[41] As plants farther away replaced older stations in populated areas, fewer consumers witnessed the smoke of power generation, which in turn added to the invisible quality of electrical power. While the statistics are incomplete for the years prior to 1920, the number of generation plants in the country fell from 2,422 in 1920 to 1,600 by 1930, a trend that reflected the move to large regional power plants that continued into the latter part of the twentieth century.[42]

Despite the reality that the capacity of coal-fired plants was increasing across the country, the myth of Niagara—that waterpower would become the main source of the growing demand for electrical power—persisted. A 1912 census report noted:

> One of the most important matters affecting the electrical industries is the use of water as a primary power. The development in electrical appliances for converting water power into electric energy, which by transmission lines is made available over large areas, together with the economies of production, indicates a continued increase in this form of primary power and a probability that it will to a still greater extent take the place of primary power derived by the use of fuel.[43]

The forecast that waterpower would outpace coal-based steam for power generation was not totally unreasonable at the time, as waterpower's share of total electric production had increased from 25 percent in 1907 to about 35 percent in 1912.[44] This trend did not continue, as by 1930 waterpower supplied less than 30 percent of the country's electrical generation, and this number declined further, making the years between 1912 and 1920 the peak years for hydro-derived electricity in the United States.[45] While many may have believed that waterpower would replace coal for the generation of electricity, plants like Fisk Street consumed 2.5 tons of coal and 1,700 tons of water, and discharged 62 tons of "waste gases and heated air" from its five smokestacks, every *minute*—and this was a plant that the *Chicago Tribune* heralded as "smokeless."[46]

The dissonance in the fact that a plant such as Fisk Street could consume two tons of coal a minute while being deemed smokeless was the result of another development that occurred simultaneously to the adoption of steam turbine generation.

Automatic stokers, which carefully regulated the draft and smoke of burning coal in power plants, eliminated observable smoke, and in turn converted the airborne effluent of combustion into an invisible state.[47] "Perfect combustion means smokeless combustion" was the mantra of power plant managers as they deployed stokers to eliminate visible smoke and raise the efficiency of coal plants.[48] Like the myth of hydropower, the apocryphal nature of the smokeless coal plant took on a life of its own. A steam plant planned for Saint Paul was noted for being "a smokeless plant to which the health commissioner may point with pride. The five big boilers are equipped with special furnaces, and the competency of the stoking equipment and the facilities for handling ashes and cinders may be best appreciated by consideration of the fact that one foreman will handle it all."[49] Engineers promoted the notion of smokeless coal even more.[50] A number of experts in the field reported that "there is no longer any necessity of polluting American cities with volumes of smoke . . . experts assert that they are operating smokeless plants and making steam economically with a coal heretofore regarded as refuse and delivered to stations for 88 cents per ton."[51] While individual power plant operators were quick to point out that their plants were not public nuisances, the federal government reinforced the public's notion that coal was harmless if burned properly. In 1913, Samuel Flagg, an engineer with the US Bureau of Mines, declared, "Coal can be burned smokeless, if you give coal the proper chance to burn."[52] Authors of a 1909 government publication, *The Smokeless Combustion of Coal in Boiler Plants*, took the position that the use of coal could be harmless and would make possible a "clean and comfortable city," and thanked the Peabody Coal Company, the Westinghouse Machine Company, and the Underfeed Stoker Company of America for providing illustrations in the book.[53]

Stokers and so-called smokeless plants did not eliminate greenhouse gases.[54] The use of stokers and turbine power plants at the same time was not part of a conspiracy to hide smoke from the public at the turn of the century. Knowledge of greenhouse gases, mercury, and other pollutants at the time was mostly limited to what people could observe, which was black smoke. The goal for engineers was simply to make black smoke go away; stokers were one way of achieving the goal, along with recommendations that coal-consuming plants be removed from areas of congested populations.[55] Between the use of stokers and the new capability of high-capacity steam turbines and alternating current transmission systems, the electrical infrastructure began to become less visible. Coal-generated electricity became more abstract and less salient with the reduction of visible smoke. Experts in the field, engineers, and the press shaped public opinion by characterizing new plants as smoke-free. The disassociation of coal from electricity increased with the removal of visible smoke from power plants in the beginning of the twentieth century. At the same time that power plants and transmission systems were becoming part of the landscape, consumer demand was expanding exponentially. The economies of scale realized by power companies led to less expensive electricity to consumers—a calculated incentive on the part of producers to maximize generation capacity—which led to more widespread adoption of electrical power. At the same time, a rising progressive middle class saw electrification as a gateway to modernity, reinforced by the newly developed art of advertising and consumer credit, which united to create an electrified consumerism, eventually leading Americans to become the highest-energy-consuming society in the world.[56]

William Leach's 1993 cultural history, *Land of Desire: Merchants, Power and the Rise of a New American Culture*, traces the formation of consumer culture in the nation primarily between the years 1890 and 1930. Showing how entrepreneurs, manufacturers, bankers, clergymen, and government leaders produced a culture of consumers, Leach concludes that the consumer capitalism that developed produced "a culture almost violently hostile to the past and to tradition, a future-oriented culture of desire that confused the good life with goods."[57] From a standpoint of periodization, paralleling Leach's study was the widespread electrification among the rising urban middle class in the country, to whom, Leach observed, "electric light was the radiant core of the consumer revolution."[58] In Leach's analysis, inexpensive electricity, first pioneered by Insull in Chicago, "disproved the widely held claim that such light would remain a luxury."[59] The effect of cheaper electricity was felt not only in Chicago; a report on a Nebraskan housewife published in the 1920 *National Electric Light Association Bulletin* provides evidence of how new economies of scale affected consumers.

> A Nebraska housewife has sent to the Nebraska Committee on Public Utility Information a graphic comparison of the cost of necessities in the home covering the period of her married life. In February, 1898, twenty-two years ago, she bought for a company Sunday dinner, six pounds of potatoes, a thirteen-pound turkey, three pounds of coffee, some raisins, mince meat, olives, celery, two pounds of tomatoes, a pound of rice and a can of asparagus. Her bill showed that this particular grocery order cost $4.74 in 1898, but that on the same date in February of this year it cost $10.82 to duplicate the order. Upon request she looked up her electric light bill of twenty-two years ago and found that she paid $11.40, as against $2.09 in the same month of 1920, and she adds, "There is no comparison in the quality of electric service now and then, when we used the old yellow light globes and knew nothing about electric irons, percolators or washing machines."[60]

In the woman's account, it can be observed that not only did the price of electricity drop considerably, electric appliances began to play a major role in the home.

Consumerism in the United States was only possible because of electrification, from electric lighting in the new class of department stores such as Wanamaker's in Philadelphia and Marshall Field's in Chicago to electric signage to the light bulbs, lamps, and household appliances now run by invisible power transmitted by wire. However inextricably tied together consumerism and electrification were, the relationship between the inconsequential consumption of electricity and the waning years of the Progressive Era runs deeper than merely contrivances and brightly lit salesmanship. David Noble, in *America by Design*, ties together the practical application of technology and capitalism in the era and demonstrates how the electric industry was "the vanguard" of not only science-based industrial development in the country but also a growing confidence in engineers and technology.[61] Electrification was the prototypical Progressive technology; it fed into the orderliness of networks and was a rational, technical solution that represented an improvement in society, especially when contrasted to fire and coal. As an advanced technology, starkly different from past energy sources, electricity meshed well with the Progressive propensity toward technical solutions to problems such as smoke. As a building block of consumerism, electrification helped to raise the standard of living across a broad swath of social classes, from J. P. Morgan to the housewife in Nebraska.

Cultural messages derived from the press, in advertising and literature, celebrated the benefits and convenience of electrical appliances and other contrivances in the home. Hoping to take advantage of consumer desires and promote electricity consumption, Insull opened the "Electric Shop" in Chicago in 1909. Catering mostly to the well off, Insull's store was devoted to the sale of electrical appliances. The shop sold a variety of appliances, including toasters, corn poppers, curling irons, heating pads, and gadgets claiming to have medical powers.[62] For those in the power production business, the forming of a new group of electrified consumers did not go unnoticed. In a 1906 issue of *Cassier's Magazine*, which catered to engineers, an article offers plant owners suggestions on how to increase central station business by stimulating demand. The author used examples such as advertising campaigns and letters soliciting the sale of electrical appliances, porch lights, and other power-consuming conveniences.[63]

As electric utilities began to team up with Madison Avenue and the rapidly refining art of advertising, the pitching of electrical appliances became more provocative, especially as it focused on women, the home, and family. In 1925 a newspaper advertisement from Insull's Commonwealth Edison calling for electrical mechanization of the home asked, "How Long Should a Wife Live?"[64] In a *Good Housekeeping* article of 1918, "We Recommend Electricity," the author promoted the use of a plug adapter that would allow for "a portable lamp, a chafing dish, toaster, percolator, and a fourth plug for the occasional use of an iron."[65] Electric consumerism was now in full swing, as women's magazines promoted more appliances and more outlets. Historian Ruth Swartz Cowan has theorized that by advertising to women, appliance sellers and power companies industrialized the household and shifted the burden of domestic work from adult men to mothers. Janice Williams Rutherford argues that electrification allowed women to be more efficient and preserved the virtuous home for women who wanted to succeed outside of the household.[66] The notion of gaining efficiency through electrical appliances inside the home ran concurrent to ideas about scientific management techniques closely tied to Progressivism. While advertising that promoted appliance sales and electricity consumption to women has been the subject of several historical studies such as Cowan's and Rutherford's, men were encouraged to buy electrical contrivances and consume power as well.

In a bulletin in 1920 from the National Electric Light Association, the electric industry's professional trade association, member companies were encouraged to advertise in a wide variety of magazines. Included in the bulletin's targeted list were magazines for boys and men along with most major women's magazines.[67] *Boys' Life* articles featured the latest discoveries in electronics such as the precursor to the television in 1920, and in *Popular Mechanics* during the same year a wide variety of products were featured, from electric bathtub heaters to arc welders.[68] Men's magazines featured a substantial number of electrical health-related devices as well, such as the "Vi-Rex Violet Ray" generator, which promised to, when rubbed on the body, cure asthma, headache, and neuralgia and restore "energy and vim."[69] The promise of electricity's potential curative powers was not just a popular phenomenon but was reflected in professional literature as well. Medical professionals promoted electricity as a therapeutic agent that could provide a cure for maladies ranging from impotence to tuberculosis. In 1907 Dr. Samuel Monell of New York published a work with the rather lengthy title *Electricity in Health and Disease; A Treatise of Au-*

thentic *Facts for General Readers, in Which Is Shown How Electric Currents Are Made to Act as Curative Remedies, Together with an Account of the Principal Diseases Which Are Benefited by Them*, which addressed the curative ability of electricity applicable in afflictions ranging from cholera to "the impaired voices of speakers and singers."[70] The extent of Monell's confidence in electricity as a vital component of a physician's toolkit is evident in his work's preface:

> The general public for the most part knows electricity simply as something for light, power, and commercial use. The Electrical Engineer knows his currents in phases, cycles, volts, and terms of copper. He works with mathematics and metals, neither of which have nerves. But the physician who prescribes electricity must know his currents in terms of tissues that feel, that breathe, and that work constantly in their wonderful ways to carry on the processes of life. To make these living tissues—nerves, blood-vessels, muscle fibres, secreting and excreting organs, heart, lungs, liver and kidneys—obey the laws of Nature and maintain health the trained physician whose curative resources are up-to-date must know electricity as the Artist knows his tools and what he can make them do.[71]

In 1919, Drs. George and Ralph Jacoby of Philadelphia published *Electricity in Medicine: A Practical Exposition of the Methods and Use of Electricity in the Treatment of Disease, Comprising Electrophysics, Apparatus, Electrophysiology and Electropathology, Electrodiagnosis and Electroprognosis, General Electrotherapeutics and Special Electrotherapeutics*, which showed equal promise in the therapeutic qualities of electricity in maladies ranging from psychoses to "paralysis of rectum."[72] While the effectiveness of various electric health devices and treatments was questionable, the claims reflect the fact that electricity itself was a new and unique force. In the early 1900s, companies advertised curative electric belts in a number of national magazines and sold them through popular catalogs such as Sears, Roebuck and Company's. Although the middle class was the intended market for the belts, historian Carolyn

Thomas de la Peña has suggested that working-class Americans purchased electric belts as well. Although many did not have access to electricity, Thomas de la Peña's analysis indicates that electricity as a technology garnered a special status in American culture.[73]

With the claims that electricity was a curative force for a variety of health conditions, the technology had now reached an omnipotent status. Defined in terms ranging from divine to a solution for social issues since 1876, electrical energy was constructed as a cure-all for ill health and an agent for convenience by the 1920s. Consumption of electricity grew rapidly as middle-class urban homes became electrified in the 1920s. The percentage of urban homes with electricity doubled between 1920 and 1930, reaching 84 percent. Radios, refrigerators, vacuum cleaners, and other electrical devices drove appliance sales from $833 million in 1921 to $1.6 billion in 1927.[74] Adjusted for per-dwelling usage, consumption rose from 339 kilowatt-hours per dwelling in 1920 to 547 kilowatt-hours in 1930.[75]

Sounding hauntingly like Mrs. Leete from Edward Bellamy's *Looking Backward*, written a quarter of a century earlier, Helen Bartlett, a teacher of cooking at Salt Lake High School in Utah, wrote in 1914, "Electricity does not poison the air. A candle uses up almost as much air as a person. Compare a candle with the burners of a gas stove and think what they do to the air."[76] A Chicago School of Sanitary Instruction noted in that same year, "When we sweep away dirty coals in favor of clean electricity for running factories, heating houses, and cooking food, we are likely to sweep away half the ills that human flesh is heir to at the same time."[77] As part of the proceedings during the National Conference of Social Work in 1926, one speaker boldly announced, "The future belongs to clean electricity and the mind can think as cleanly. We shall make little progress with programs until we learn to include electricity and the mind of the youth in these programs."[78] As electricity was increasingly associated with the notion of clean—"clean electric motors," "clean electric plants," "clean electric light"—it was increasingly evident that the association between electricity and coal, fire, and smoke had been severed in the mind of society.[79] While Americans celebrated "clean electricity," and "the sweeping away of dirty coals," electric utility companies' coal consumption rose from 31,640,000 tons in 1920 to 40,278,000 tons in 1930.[80]

CONCLUSION

Journalist Frank Bohn's feature "The Electric Age: A New Utopia," in the October 2, 1927, issue of the *New York Times Magazine*, reflects a culmination of social messages that had been attached to electrification:

> Our world is being torn down and rebuilt. Economic society is now experiencing the most remarkable transformation in its history. The last century and a half constitutes a peculiar epoch: It began with the steam engine and the first industrial revolution. It is now closed by the electric superpower system and the new industrial revolution. Looking back upon it we now begin to see it in clear historical perspective. The older industrialism should be seen as the introduction to the new. The first industrialism was the gift of England. The new revolution is the peculiar creation of this country. Two-thirds of the machines in our American factories are now run by power from central plants. We have already entered the Electric Age. The new process is changing much more than the mere means of production. It is transforming the inner as well as the outer conditions of human life. We see, we hear, we speak, we learn, by means of new and marvelous mechanical instruments. These instruments have already begun to work their change upon the individual mind and society. Yet we are only beginning.[1]

Richly adorned with a photograph of Niagara Falls, the cover story sent a strong message that electricity was a nonsteam, noncoal, energy panacea. Bound together with themes of transformation, electricity was portrayed as a complete departure from all that had come before. Through the declaration that the country had now entered the electric age, the acquisition of more electrical appliances was encouraged as a gateway to an improved future. Absent from Bohn's article was any mention of coal. With Niagara as the featured photograph and the statement that the age of steam had passed, the true nature of electrification was obscured behind the mist of falling water. In the eyes of Bohn and other Americans who saw electricity as a cure for society's ills, electrical energy was changing the means of production, changing individuals, and changing society.

Bohn was not alone in his electrical boosterism. Newspapers and magazines in the opening decades of the century celebrated the miracle of electricity, as did those in society who consumed it. In an edition of the *Washington Times* from 1922, a Mrs. W. H. Stewart won a letter-writing contest with this entry:

> Electrical experience will convince any woman that a few yards of electric wire in her home can do more toward her emancipation than Congress can. Gone is blue Monday. The electric washer has transformed it into roseate hue. An interesting hour in the morning results in the completion of the family wash. Afternoon free for study or profitable shopping. Ironing, too, has resolved itself into a pleasure, when an electric iron guards off overheating. Because of electricity, the servant question has been solved. A vacuum, not woman, does the cleaning, eliminating dusting, unsightly hands and dust-filled pores. There's now time for woman's mental improvement, social pleasure and electrical massage, which stays the marks of time. Breakfast grouchiness, so trying to every stage of married life, is vanquished by electricity. Now, mother, sweet and beautifully coiffured (her electric curler has superseded curl papers) makes no tiresome trips into the kitchen, but serves coffee from an electric percolator, toast from an electric toaster, and who can enumerate the advantages of the grill? Rested and smiling when husband comes home, wife plays their favorite music on the electric graphophone or piano. No club

for him now. Together they sit beneath the softly shaded electric lamps, happy as lovers.[2]

During the same decade in which Bohn and Stewart wrote, the percentage of urban homes with electricity began to increase, from 47 percent in 1920 to 84 percent in 1930. For some working-class people and tenement dwellers in the first two decades of the 1900s, electricity in the household was a marker of modernity and propriety, as it represented an alternative to the drudgery of managing fires and fuel.[3] Whether one lived in an electrified household or not, the enthusiasm for clean, convenient energy cut across class lines. By the close of the 1920s, the "typical" house has not only electricity but also several appliances.[4] By 1921, 64 percent of urban, "industrial class" homes has at least an electrical iron, and over 40 percent of "average" homes had a vacuum cleaner.[5] Clearly, electricity had met an enthusiastic acceptance as a necessary part of American life by 1930.

While journalists and the lay public had nothing but praise for the benefits of electrical power, cultural messages from experts and those influential in society continued to position electricity as environmentally inconsequential. Charles Robert Gibson, a distinguished associate of the Institute of Electrical Engineers, wrote in his book *The Romance of Modern Electricity*, "It seems to me very probable that before another generation has come and gone people will have no cause to grumble at smoke and dirt in the atmosphere of cities, as the whole energy required for motive power, heating, and lighting may be delivered from one great generating station outside of the city."[6] Gibson's belief that moving generation stations outside of the city was a solution to smoke and dirt in the atmosphere reflects a mentality perpetuated by electricity's unique ability to travel long distances by wire. His supposition also demonstrates how the environmental aspect of energy exceptionalism had manifested itself, that the atmosphere was both forgiving and too big to foul. Just as engineers in the 1870s had advocated higher chimneys to solve the problem of smoke, the out of sight, out of mind mentality remained firmly entrenched in the 1930s.

Writing in 1934, influential architect, historian, and social theorist Lewis Mumford reinforced the theme that electrification marked the beginning of a new age, which he referred to as the "neotechnic phase" of civilization. In his work *Technics and Civilization*, Mumford wrote,

> Unlike coal in long distance transportation, or like steam in local distribution, electricity is much easier to transmit without heavy losses or higher costs. Wires carrying high tension alternating currents can cut across mountains which no road or vehicle can pass over; and once an electric power utility is established, the rate of deterioration is slow. Moreover, electricity is readily convertible into various forms: the motor, to do mechanical work, the electric lamp, to light, the electric radiator, to heat, the x-ray tube and the ultra-violet light, to penetrate and explore, and the selenium cell, to affect automatic control.[7]

Mumford's writing reflects just how pervasive ideas about electricity as a technological actor independent of coal were. Breaking completely from a previous period he referred to as "carboniferous capitalism" in a "paleotechnic" phase of society, Mumford's neotechnic phase of civilization marked the beginning of a separation from the burdens of coal and steam that had come before.[8] As with those in the past who saw electricity in religious terms or as a solution to social problems, Mumford's work perpetuated the utopian narrative of electricity as a panacea. Along with positioning electricity as a harbinger of modernity, cultural messages that equated it with energy abundance encouraged consumption and related electricity to American nationalism.[9] Increased usage of electrical power was becoming a point of national pride that President Herbert Hoover saw as a force that had "lifted the drudgery from the lives of women" and had "taken sweat from the backs of men."[10]

In a span of fewer than fifty years, from Edison's start-up of the Pearl Street Station in Manhattan to pundits' pronouncements that the age of steam had passed, coal-fired electrical generation in the United States increased exponentially. As electric companies were promoting their energy as a gateway to modernity, coal consumption grew from thirty-one million tons in 1920 to forty million tons a decade later. By 1940, coal consumed for the generation of electrical power would nearly double the amount consumed just twenty years earlier.[11] As central stations further separated power generation from power consumption, they also separated coal from consciousness. The energy infrastructure behind the wall outlet became obscured as cultural messages positioned electricity as a new kind of energy that was different from the age of steam that had come before. Through the entire time that this new energy consciousness was occurring, more steam turbines were coming online, and the fires from coal continued to burn.

This study ends in the 1930s. By the opening of that decade, electricity had become positioned as energy-independent from coal and a new American energy consciousness was in place. Unique from any technology that had come before it, the necessity of electricity had become such that Franklin Roosevelt would make it a cornerstone of his 1932 bid for the presidency. In his Portland speech of September of that year, Roosevelt effectively declared that electricity was an American right, a "definite necessity" that would preserve the "social order" of American life.[12] Later, when Roosevelt signed executive order number 7037 in 1935 to create the Rural Electrification Administration, he put into action legislation that would further expand the reach of electrical power. Prior to Roosevelt's directives, power companies had already begun gearing up for runs of power that would add more space between electrical generation and consumption. High-voltage lines and systems would now allow for the economical transmission of electricity to greater distances, further separating coal from the purview of American society.[13]

While the physical distance between electrical generation and consumption continued to increase beyond the 1930s, cultural messages removing coal from the equation continued as well. As the Tennessee Valley Authority (TVA) became a subject of national attention under FDR's presidency and into World War II, TVA war propaganda posters picturing dams and pronouncing "Out of Water Power Comes Air Power" continued to associate electricity exclusively with hydropower.[14]

While this may have been true for some defense plants in the South, it was by no means the rule, and by 1944 the burning of coal for electricity generation would double that of its rate in 1920.[15] Beyond the 1940s and into the era of nuclear electrical generation beginning in the 1950s, rhetoric such as that contained in President Eisenhower's "Atoms for Peace" speech in 1953 and Allan C. Fisher Jr.'s article "You and the Obedient Atom" in *National Geographic* in 1958 reinforced ideas that electricity was and would be in the future separate from coal. When Eisenhower asserted that "abundant electrical energy in the power-starved areas of the world" would be provided by the atom, he perpetuated ideas about a coal-free future of endless energy.[16] Five years later, Fisher promised "abundant energy released from the hearts of atoms" to a generation unaware of the circumstances of their own power consumption.[17]

While the contribution of damaging environmental effects from nearly eighty years of coal-fired electrical generation were largely unknown to the public in the 1960s, the salient dangers of nuclear power manifested themselves in stark contrast to the invisible threat of burning coal. The very real consequences of early nuclear disasters such as that at Three Mile Island in 1979 went far to shift the public's attention further away from the deleterious effects of burning coal that would begin to come to light in the same decade.

From today's perspective, it is not easy to imagine the world without electricity. This book argues that the nature of electrification as an energy system contributed to a culture of American energy and environmental exceptionalism, but it does not contend that electricity was in itself a technology of declension. The fast adoption of electricity was clear evidence of its acceptance by and usefulness to society. In his work of 1988, *The Evolution of Technology*, historian George Basalla explores the idea that technology evolves to meet the needs of humanity.[18] While Basalla asserts that scholars have maintained that technology is not necessary for meeting animal needs in humans, humanity began to cultivate technology for "human life, the good life, or well-being."[19] Regardless of one's position on the issue of technology as a necessity or as production of the superfluous,

Americans adopted a culture in which electricity was sine qua non, an essential element for well-being. In the process of doing so, the primary method of electrical generation was dictated by the social demand and available technology.

Just as this book does not contend that electricity was a regrettable technological advance, it also does not contend that electrical infrastructure in the United States could have evolved differently than it did. The only viable mechanical rotary prime movers that existed during Siemen's and Gramme's perfection of practical dynamos were hydropower and steam engines. Considering the capital cost, mobility, and on-demand capability of the steam engine, it was the clear choice for generating electricity. The low cost, efficiency, and availability of coal dictated that it would be the primary energy source for the evolving electrical grid. American society's demand for electricity—in large part created by the utility companies who would profit from it—consistently outstripped any realistic chance of a practical renewable or alternate energy electrical infrastructure. In the first century of electrification, forces of consumerism and convenience, along with the cultural positioning of electricity as the gateway technology to modernity, combined with the invisible nature of generation and distribution to eliminate any social need for noncoal alternatives.

Revisiting those who did pursue alternative energy technologies in American history reveals a tragedy of timing. At every juncture, alternatives to coal were impractical for supplying the around-the-clock demand dictated by a rapidly growing nation and there were no pressing social trends to drive the development of sustainable sources of energy. Coal was inexpensive, had technological momentum, and the long-term effects on the environment were unknown.

Prior to the development and adoption of electricity as a practical power source, utopianist John Adolphus Etzler was pursuing wind power as an alternative to coal. Etzler claimed that there were "powers in nature at the disposal of man, million times greater than all men on earth could effect."[20] Etzler was studying engineering in 1824 at the same time that French mathematician Nicolas-Leonard-Sadi Carnot published *Reflec-*

tions on the Motive Power of Heat. Carnot surmised that "heat [was] simply motive power which has changed form," and that air currents should be studied to "their smallest details" in the pursuit of means for motive power.[21] Carnot's ideas that heat from the sun continually agitated the atmosphere were at the core of Etzler's energy pursuits, along with his anticapitalist visions of communal energy. Etzler's issue with coal was not in response to its environmental effects, but with the capitalists who controlled the supply and the price. In one of the more prescient statements in the history of energy, Etzler referred to industrialism as nothing more than a "vicious energy monopoly."[22] While he did pursue and patent two energy-related inventions, he was unable to find financial support or develop a scalable wind engine at a time when the United States was rapidly expanding. Although Etzler was working on noncoal motive power before electrification, his pursuit of renewable energy sources provides an intriguing counterfactual history, yet also exposes the issue of viability. From Etzler forward, those in pursuit of noncoal systems found that energy demand outpaced the practicality of alternative means for generation. Etzler was not alone in his belief in alternative sources of energy. In 1852, Abraham Lincoln lectured on "Discoveries and Inventions" and expressed a similar sentiment for the potential of wind power:

> Of all the forces of nature, I should think the *wind* contains the largest amount of *motive power*—that is, power to move things. Take any given space of the earth's surface—for instance, Illinois—and all the power exerted by all the men, and beasts, and running-water, and steam, over and upon it, shall not equal the one hundredth part of what is exerted by the blowing of the wind over and upon the same space. And yet it has not, so far in the world's history, become proportionably *valuable* as a motive power. It is applied extensively, and advantageously, to sail-vessels in navigation. Add to this a few wind-mills, and pumps, and you have about all. That, as yet, no very successful mode of *controlling*, and *directing* the wind, has been discovered; and that, naturally, it moves by fits and starts—now so gently as to scarcely stir a leaf, and now so roughly as to level a forest—doubtless have been the insurmount-

able difficulties. As yet, the wind is an *untamed*, and *unharnessed* force; and quite possibly one of the greatest discoveries hereafter to be made, will be the taming, and harnessing of the wind.[23]

As with Etzler and Lincoln, John Ericsson saw the future in renewable energy and wrote of a future without coal. In 1851, Ericsson sailed the *Ericsson*, a ship he designed and which was powered by a caloric heat engine in trials, but considered it a failure because it did not match the speed of coal-fired vessels. Ericsson continued to pursue a practical caloric engine throughout his life, and in 1876 he was denied in his application to be an exhibitor at the Centennial Exhibition where Edison, Bell, and others were demonstrating their inventions. Within Ericsson's graduate thesis, "The Use of Solar Heat as a Mechanical Motor-Power," of 1868 from the University of Lund in Sweden, he wrote:

> I cannot omit adverting to the insignificance of the dynamic energy which the entire exhaustion of our coal fields would produce, compared with the incalculable amount of force at our command, if we avail ourselves of the concentrated heat of the solar rays. Already Englishmen have estimated the near approach of the time when the supply of coal will end, although their mines, so to speak, have just been opened. A couple of thousand years dropped in the ocean of time will completely exhaust the coal fields of Europe, unless, in the meantime, the sun can be employed.[24]

If Ericsson had been granted approval to demonstrate his solar engine at the exhibition in 1876 in the presence of men such as Thomas Edison, William Wallace, and Dr. Joseph Henry, would they have had an interest in pursuing this technology? Considering that Ericsson's engine was by no means perfected or practical for the continuous demands of electrical generation, the answer is likely no. It was the Corliss steam engine that captured the imagination of most of the attendees at the exhibition, rendering any potential interest in a caloric heat engine unlikely. While Ericsson's work on a viable heat engine did not come to fruition, he is credited with the invention of the "Ericsson Cycle," which identified a thermodynamic process with a regeneration stage that boosts efficiencies in the operation of turbine engines. In

1979 a paper identified the Ericsson Cycle as capable of "unprecedented efficiency" when applied to power plant turbines.[25]

Nikola Tesla's early financial backers, Alfred Brown and Charles Peck, were interested in a system for generating steam based upon the principles embodied in the cryophorus, a device developed by the English scientist W. H. Wollaston in 1805. The cryophorus utilized temperature differentials present in the ocean to generate steam, yet the more urgent pursuit for Brown, Peck, and Tesla in 1886 was the creation of an alternating current electric motor that would fund more development work.[26] Although Tesla's work on the cryophorus was put on hold as he continued his career with Edison and Westinghouse, he did revisit the idea in 1931 in an article titled "Our Future Motive Power," in which he explored several concepts for a workable temperature differential generation system. The pragmatic Tesla concluded that through all of his conceptual designs of a cryophorus-based system, the "performance [was] too small to enable successful competition with present methods."[27] In the same article, Tesla expressed his belief that photoelectric cells would be "of practical importance in the future," and that "our stores of coal and oil will be eventually used up." Although Tesla's ideas for harnessing ocean thermal energy conversion (OTEC) did not come to fruition during his lifetime, research into his concepts continued and a 105-kilowatt OTEC plant went online in Hawaii in 2015.[28]

While Ericsson and Tesla pursued solar power, Charles Brush, who controlled the arc light market in the late 1870s, developed a windmill that generated electricity behind his mansion on Euclid Avenue in Cleveland. The Brush windmill powered a dynamo that provided power for his home and stored electric energy in batteries for use when the days were calm. For all of his genius and foresight, pundits saw the windmill as too costly to operate when compared to coal.[29] As with Etzler and Ericsson, Brush was pursuing a technology without a market need nor the capacity to generate electricity on a large scale at the time. Knowing that the project lacked viability, Brush never bothered to patent his wind system. To quote Robert W. Righter's *Wind Energy in America: A History*, "the Brush windmill represented a nineteenth-century concept that could not fit into twentieth-century complexity and interdependence."[30] The premature nature of wind power technology also befell rancher Frank Bosler outside of Laramie, Wyoming, whose inquiry into pursuing wind power in 1912 was quickly dismissed by General Electric, as the company was pursuing the development of large steam turbines for customers such as Samuel Insull in Chicago.[31]

It would be easy to blame capitalism in general and the coal and utility interests specifically for not pursuing these and other alternative energy technologies, but it would also be an exercise in historical presentism. Even though energy visionaries such as Etzler, Ericsson, and Tesla saw the future as one of resource depletion, successful energy technologies were defined by viability, and there was no economic incentive for a robust pursuit of renewable systems. In addition to no economic incentive, there was no social incentive for pursuing alternate energy systems. Lacking a connection between coal and electricity and coal and atmospheric decline, the status quo of coal has only recently become a social issue. Amid heightened attention on smoke from industrial applications and a growing number of internal combustion engines, there is no compelling evidence that smoke from electrical generation was a public concern until much later. Electricity conceptualized as the postcoal energy source was a social construct that was firmly in place in the early 1900s, and many more obvious forms of air degradation existed at the time.

Even though coal-burning central power generation stations were responsible for a significant portion of atmospheric fouling during the first seventy-five years of the twentieth century, they were not specifically implicated until the waning decades. Although scientists including John Tyndall and Svante Arrhenius discovered a theoretical link between carbon dioxide from fossil fuels and climate change in the last half of the nineteenth century, these were studies confined to the laboratory and were not social concerns.[32] By the 1930s in the United States, there were articles in the popular press claiming that winters had gotten warmer, but no one appeared to be worried about the changes. Meteorologists at the time correctly explained that weather patterns were cyclical and varied over time.[33] In 1938, Guy Stewart Callendar stood before the Royal Meteorological Society in London and claimed that human industry was changing the climate, but few scientists acknowledged his claims, and there is no evidence that US scientists were interested, and certainly not the American public. By the mid-1950s, air pollution was not yet a serious subject of public policy, with the US federal government offering a limited program of technical assistance only. Confined to a few large cities, air quality control programs existed during a time when the problem was attributed primarily to coal combustion, but not broken out by industry.[34] Subsequent federal legislation in 1963 and 1967 enabled the government to begin investigating and monitoring air quality, including stationary sources such as factories and power plants, and in 1970 the Clean Air Act (CAA) was greatly expanded. Although electrical generation plants were responsible for about 40 percent of carbon dioxide emissions and most of the sulfur dioxide emitted into the atmosphere, public perceptions of the principal sources of air pollution focused mostly on automobiles.[35] This focus was clearly justified, as leaded gasoline and carbon dioxide emissions from vehicles were a public health hazard, yet with attention mostly on automobiles, stationary sources such as power generation plants slipped further from public view.

In 1971, plant ecologists Dr. F. Herbert Bormann and Dr. Gene Likens were conducting research in the White Mountains of New Hampshire and discovered that the acidity of water throughout the East Coast had increased 1,000 percent since the 1950s.[36] Working in some of the same forests that the first colonists had once marveled over as potential sources of fuel, Bormann and Likens found that the pH level of the rain and snow was also of abnormally high acidity.[37] Over the next several years, the scientists traced the acidity to sulfur dioxide and nitrogen oxides emitted from smokestacks, many of which are the byproducts of steam turbine coal plants that produce electricity. Bormann and Liken's discovery did place the term "acid rain" into the public vernacular, and further implicated the generation of electricity as a prime culprit, but the impact of the discovery on the public opinion is unclear. Prior to the year 2000, there are little survey data available on public attitudes toward energy.[38]

The public was not aware of the detrimental effects of coal-fired power generation in the first century of electrification in the United States. When components of environmental degradation such as smog, air pollution, and acid rain did become social issues in the 1970s, the effects of coal-fired generation plants were lost among a myriad of other sources. Just as electricity itself lost its salience, the environmental effects caused by its generation were subtle and outside of the consciousness of the consuming public. Up to the 1970s, embedded cultural perceptions of energy exceptionalism and environmental inconsequentiality held strong.

In 1973, for the first time in the nation's history, how energy was conceptualized by society abruptly changed. During the first oil crisis, gasoline shortages shook the very foundation of American confidence by demonstrating that the United States had lost its long-held energy independence. Not only was oil in short supply but the cost of electricity had tripled as well. For forty years before 1973, the cost of electricity in the United States had been stable, and in some cases the cost of electricity had declined relative to the consumer price index. After 1973, the increased cost of oil caused regional electricity prices to soar due to the cost of generation in regions of the country that depended on oil.[39] Reminiscent of Etzler and Ericsson more than a century earlier, President Jimmy Carter declared in 1978 that "acres of mirrors [can] generate steam for electricity," hinting at an electric and alternate solution to the nation's sudden energy crisis.[40] A year later, Carter had thirty-two solar panels installed on the roof of the White House to heat water. Even though engineers claimed that the solar panels were working as intended, in 1986, President Ronald Reagan's administration quietly removed them.[41]

The Reagan administration's dismantling of the White House's solar panels was symbolic of the United States's lingering defiance of energy vulnerability, and reflected a stubbornness to let go of the nation's fossil fuel status quo. In the two decades from 1980 to the year 2000, published reports on the effects of burning fossil fuels began to emerge with regularity in the press. Of the some 1,200 global-warming-related articles published in the *New York Times* in those two decades, over one hundred articles implicated coal-driven electrical generation plants as culpable in atmospheric change. Although the press was starting to link coal-fired electrical generation to climate change, Americans seem to have met the news with indifference.[42] This nonchalant attitude is in part informed by a long-held view of electrical energy that is modern and poststeam.

While American attitudes about major social issues such as race and gender have undergone significant shifts since the beginning of the twentieth century, ideas about energy consumption in general and electricity specifically have been dangerously enduring. As late as 2005, Americans still believed that hydropower was the nation's top energy source when in reality it accounted for less than 10 percent.[43] Seven years later, in 2012, after much press exposure on the link between coal and climate change, only 38 percent of Americans identified burning coal as the source of most electricity in the United States.[44] In interviews of approximately 350 college students between 2000 and 2015, fewer than 20 percent made a connection between their cell phones and coal-derived power.[45] These latest findings may suggest that battery-powered devices and vehicles further break the link between coal, electricity, and environment. While part of this energy illiteracy is due to the nonsalient nature of the country's infrastructure, another part is derived from a deep connection between past and present.

From its very beginning, electricity was a technology of myth. Images of Franklin with his kite and Edison the wizard morphed into Bohn's Niagara-fueled poststeam society. Beginning in the last decades of the nineteenth century, the simultaneous cultural and physical separation of coal from electricity reinforced long-held American attitudes about energy and the environment. In the process, a new sense of energy exceptionalism without environmental consequences was formed. American society still perceives an imagined energy environment, due to a lack of smoke and smokestacks and social views about energy that began between the 1870s and 1930s. Current beliefs that new technologies, such as plug-in electric automobiles, are inconsequential environmental saviors are remnants of ideas first promoted in the early twentieth century. The artifacts of inconsequential electrical consumerism still inform society's predilection for increasingly high-tech cell phones, computers, and other electronic devices that now make up nearly 10 percent of an average US household's electricity consumption.[46]

Since the arrival of the first Europeans in North America, the New World was defined as a place of unlimited energy coupled with a clean environment. As smoke began to turn the nation into a clone of the Old World, electric light and technologies of wire offered a clean contrast to what had come before. Electricity was the first technology to separate fire from usable power via wire, and from its earliest applications, the unique spatial dynamics of electricity led to an abstraction of energy. As the origins and consequentiality of electricity became increasingly blurred, society came to see the force as magical, mysterious, and transformative, and aspired to build a nation based on the imagery of the White City, where Henry Adams's "dirty engine house" was no longer in view. While cultural messages reinforced the inconsequentiality of electricity through the promotion of waterpower and smoke-free technologies, a growing demand for inexpensive electricity ultimately led to the building of more coal-based generating plants. As the twentieth century began, technical and cultural forces converged to further abstract energy use. These forces in turn led to more consumption, setting the stage for Americans to become the number one producer of carbon dioxide emissions in the world, with approximately 40 percent of the emissions from the generation of electricity.[47]

The purpose of this study has been to show how the evolution and adoption of electricity in the United States was a factor in the formation of an ideology of inconsequential energy consumption. By its very nature, as a secondary energy source that could be transmitted inconspicuously via wire and expanded technical systems, the adoption of electricity led to a culture whose energy literacy stopped at the light switch or the wall outlet. Americans confronted electrification as an essential force and defined it as a modern technology with no limitations or consequences. As it became adopted, electricity evolved into a technology of consumption that was ubiquitous and always on, with origins that were misconstrued and misunderstood. Those who deny climate change and pursue more fire-based energy are part of a historical legacy.

On a poster designed for the first Earth Day in 1970, *Pogo* cartoonist Walt Kelly famously wrote, "We have met the enemy, and he is us."[48] The poster conveyed the point that destruction of the environment is not due solely to uncaring profiteers or producers, and expressed the idiosyncrasies of the human condition. Individuals in society perpetuate demand for goods and services and share the blame for overconsumption and environmental exploitation.

The first generation of electric consumers established a social definition of electricity as an inconsequential energy technology, and the environmental effects that resulted are now too big to ignore. Plausible deniability as a justification for the unfettered use of inexpensive energy is no longer a tenable position. Adam Frank, an astrophysicist, recently proclaimed that "climate change is not our fault." In explanation, Frank asserts that the factors contributing to climate change were not understood when most of the damage occurred.[49] As society lags behind science in the acceptance of climate change, and policymakers lag behind society, recent policy actions further limiting central generation stations' coal emissions are encouraging, yet are still experiencing pushback from the coal industry. Despite a seemingly broader understanding of the link between coal and the burning of other fossil fuels and climate change, the direct connection between electricity, coal, and fuel sources such as hydraulically fractured natural gas are still unclear to many energy consumers. As we move into an age of electrical energy that is increasingly battery-powered, the connection between power generation and power consumption becomes even more removed. Much of the energy powering plug-in automobiles and appendage-like smartphones still derives from coal-fired steam technology from the nineteenth century. We can only hope that through the teaching of energy awareness we can reveal the obvious consequences of our first century with electricity.

NOTES

INTRODUCTION

1. Frank Bohn, "The Electric Age: A New Utopia," *New York Times*, Oct. 2, 1927.

2. Henry Adams, *The Education of Henry Adams: An Autobiography* (Boston: Houghton Mifflin, 1918), 292.

3. Matin V. Melosi, *Effluent America: Cities, Industry, Energy, and the Environment* (Pittsburgh, PA: University of Pittsburgh Press, 2001), 14.

4. Kevin Coyle, *Environmental Literacy in America: What Ten Years of NEETF/Roper Research and Related Studies Say about Environmental Literacy in the U.S.* (Washington, DC: National Environmental Education and Training Foundation, 2005), v.

5. Leo Marx, *The Machine in the Garden: Technology and the Pastoral Ideal in America* (New York: Oxford University Press, 1964), 103.

6. Marx, *Machine in the Garden*, 221.

7. David Stradling, *Smokestacks and Progressives: Environmentalists, Engineers and Air Quality in America, 1881–1951* (Baltimore, MD: Johns Hopkins University Press, 1999).

8. William Cronon, *Nature's Metropolis: Chicago and the Great West* (New York: W. W. Norton, 1992), 224.

9. Cronon, *Nature's Metropolis*.

10. Melosi, *Effluent America*.

11. Melosi, *Effluent America*, 24. See also Lewis Mumford, *The City in History: Its Origins, Its Transformations, and Its Prospects* (New York: Harcourt, Brace and World, 1961), 447.

12. Melosi, *Effluent America*, 175.

13. Thomas Parke Hughes, *Networks of Power: Electrification in Western Society, 1880–1930* (Baltimore, MD: Johns Hopkins University Press, 1983).

14. Hughes, *Networks of Power*, 280.

15. Hughes, *Networks of Power*, 201.

16. David E. Nye, *Electrifying America: Social Meanings of a New Technology, 1880–1940* (Cambridge: MIT Press, 1990).

17. Nye, *Electrifying America*, 157.

18. Robert H. Wiebe, *The Search for Order, 1877–1920* (London: Macmillan, 1967), 157.

19. David F. Noble, *America by Design: Science, Technology, and the Rise of Corporate Capitalism* (Oxford: Oxford University Press, 1979), 18.

20. Maury Klein, *The Power Makers: Steam, Electricity, and the Men Who Invented Modern America*, 1st US ed. (New York: Macmillan, 2008), 71.

21. Klein, *Power Makers*, 75.

22. Klein, *Power Makers*, 415.

23. Forrest McDonald, *Insull* (Chicago: University of Chicago Press, 1962).

24. Christopher F. Jones, *Routes of Power: Energy and Modern America* (Cambridge, MA: Harvard University Press, 2014).

25. Vaclav Smil, *Energy Transitions: History, Requirements, Prospects* (Santa Barbara, CA: Praeger, 2010).

26. Richard F. Hirsh and Benjamin K. Sovacool, "Wind Turbines and Invisible Technology: Unarticulated Reasons for Local Opposition to Wind Energy," *Technology and Culture* 54, no. 4 (2013): 705–34.

27. Richard Hofstadter, *The Age of Reform: From Bryan to F.D.R.* (New York: Vintage Books, 1955), 23.

28. "John Ericsson," *Science* 13, no. 319 (Mar. 15, 1889): 191.

29. R. J. Bufard of General Electric to Frank C. Bosler, Apr. 24, 1913, box 47, folder 403, Bosler Collection, American Heritage Center, University of Wyoming, Laramie.

CHAPTER 1: ENGLISH ROOTS, UTOPIA FOUND AND LOST

Epigraph: Edward Bellamy, *Looking Backward: 2000–1887* (Boston: Ticknor, 1888), 165.

1. Catharine Esther Beecher and Harriet Beecher Stowe, *The American Woman's Home: Or, Principles of Domestic Science; Being a Guide to the Formation and Maintenance of Economical, Healthful, Beautiful, and Christian Homes* (New York: J. B. Ford, 1869), 424.

2. David Stradling, *Smokestacks and Progressives: Environmentalists, Engineers and Air Quality in America, 1881–1951* (Baltimore, MD: Johns Hopkins University Press, 1999), 47.

3. John H. Griscom, *The Uses and Abuses of Air: Showing Its Influence in Sustaining Life, and Producing Disease; with Remarks on the Ventilation of Houses, and the Best Methods of Securing a Pure and Wholesome Atmosphere inside of Dwellings, Churches, Courtrooms, Workshops, and Buildings of All Kinds* (New York: Redfield, 1848), 75.

4. Wil Roebroeks, Paola Villa, and Erik Trinkaus, "On the Earliest Evidence for Habitual Use of Fire in Europe," *Proceedings of the National Academy of Sciences of the United States of America* 108, no. 13 (2011): 5210.

5. Pliny the Elder, *The Natural History*, vol. 1 (London: George Bell and Sons, 1893), 142.

6. Charles Darwin, *The Descent of Man and Selection in Relation to Sex*, vol. 1 (New York: D. Appleton, 1872), 176.

7. Stephen J. Pyne, *Fire: a Brief History* (Seattle: University of Washington Press, 2001), 111.

8. King Camp Gillette, *The Human Drift* (Boston: New Era, 1894).

9. Leo Marx, *The Machine in the Garden: Technology and the Pastoral Ideal in America* (New York: Oxford University Press, 1964), 25.

10. Francis Higginson, *New Englands Plantation; or, a Short and True Description of the Commodities and Discommodities of That Country* (London, 1860), 8.

11. Peter Linebaugh, *The London Hanged: Crime and Civil Society in the Eighteenth Century* (Cambridge: Cambridge University Press, 1992), 378. For the loss of estover rights as a result of enclosure, see Linebaugh, *The Magna Carta Manifesto: Liberties and Commons for All* (Berkeley: University of California Press, 2008), 134.

12. Sussex Archaeological Society, *Sussex Archaeological Collections Relating to the History and Antiquities of the County* (Lewes, England: Sussex Archaeological Society, 1848), 169. The Sussex record is just one of many that are demonstrative of wood for fuel held as a vital component of the land; see also John Perlin, *A Forest Journey: The Role of Wood in the Development of Civilization*, 1st ed. (New York: W. W. Norton, 1989).

13. Joan Thirsk, *Chapters from the Agrarian History of England and Wales, 1500–1750* (New York: Cambridge University Press, 1990).

14. Peter Brimblecombe, *The Big Smoke: A History of Air Pollution in London since Medieval Times* (London: Routledge, 2011), 9.

15. Mark Z. Jacobson, *Atmospheric Pollution: History, Science, and Regulation* (Cambridge: Cambridge University Press, 2002), 83.

16. Thomas Browne, *The Works of Sir Thomas Browne* (London: Henry G. Bohn, 1852), 340.

17. Ralph Waldo Emerson, *The Journals and Miscellaneous Notebooks of Ralph Waldo Emerson*, ed. A. W. Plumstead (Cambridge, MA: Belknap Press of Harvard University Press, 1969), 7; Ralph Waldo Emerson, *Nature* (Boston: J. Munroe, 1836). Emerson celebrates clean air in several passages in *Nature*, perhaps the most well-known quote being "Standing on the bare ground—my head bathed by the blithe air, and uplifted into infinite space, all mean egotism vanishes" (8). Emerson also connects to Harriet Beecher Stowe in that Emerson and Stowe were two of the original contributors to the *Atlantic* in 1857.

18. John Evelyn, *Fumifugium, or, the Inconveniencie of the Aer and Smoak of London Dissipated Together with Some Remedies Humbly Proposed* (London, 1661), 26.

19. Evelyn, *Fumifugium*, 15.

20. Mary Anne Everett Green et al., *Calendar of State Papers, Domestic Series, of the Reign of Charles II, 1660–1685* (London: H. M. Stationery Office, 1860), 507.

21. Higginson, *New Englands Plantation*, 11.

22. Green et al., *Calendar of State Papers*.

23. John Smith et al., *The Generall Historie of Virginia, New-England, and the Summer Isles: With the Names of the Adventurers, Planters, and Governours from Their First Beginning An.: 1584 to This Present 1626* (London, 1632), 90.

24. Smith et al., *Generall Historie*, 388.

25. Samuel Purchas, *Purchas His Pilgrimes: In Five Bookes*, vol. 3 (London, 1625), 852.

26. Louis Hennepin, Louis Joliet, and Jacques Marquette, *A New Discovery of a Vast Country in America Extending above Four Thousand Miles, between New France and New Mexico. With a Description of the Great Lakes, Cataracts, Rivers, Plants, and Animals* (London, 1698), 118.

27. William Cronon, *Changes in the Land: Indians, Colonists, and the Ecology of New England* (New York: Hill and Wang, 1983), 20.

28. Thomas Harriot, *A Briefe and True Report of the New Found Land of Virginia* (1590; repr., New York: Dover, 1972), 69, 55, 56–57, 60, 63, 66.

29. Cronon, *Changes in the Land*, 54.

30. William Dean Howells, *A Traveler from Altruria: Romance* (New York: Harper and Brothers, 1894), 129.

CHAPTER 2: THE ENERGY REVOLUTION AND THE ASCENDANCY OF COAL

Epigraph: Israel W. Morris, *The Duty on Coal; Being a Few Facts Connected with the Coal Question, Which Will Furnish Matter for Thought to the Friends of American Industry* (Philadelphia: Baird, 1872), 3.

1. Charles Grier Sellers popularized the concept of a "market revolution" in the Jacksonian Period; see *The Market Revolution: Jacksonian America, 1815–1846* (New York: Oxford University Press, 1991), 7.

2. Historian Thomas Hughes first developed the concept of "technological momentum." Hughes saw developing technological systems being influenced by social forces, such as acceptance of the systems by society. As systems mature, they become more independent of outside social forces—they gain a momentum as status quo by way of "acquired skill and knowledge, enormous physical structures, as well as special machines and processes." Coal train networks, steam engines burning coal, and the development of domestic coal stoves and furnaces are all examples of how coal gained momentum beginning in the mid-nineteenth century. For a more complete explanation, see Hughes's essay in Merritt Roe Smith and Leo Marx, eds., *Does Technology Drive History? The Dilemma of Technological Determinism* (Cambridge: MIT Press, 1994), 101. Also see W. E. Bijker et al., *The Social Construction of Technological Systems: New Directions in the Sociology and History of Technology* (Cambridge: MIT Press, 2012), 70.

3. *Somatic energy* in this usage refers to "energy of the body," or muscle power of either humans or animals; see Vaclav Smil, *Energy in World History* (Boulder, CO: Westview Press, 1994), 3. Also see John R. McNeill, *Something New under the Sun: An Environmental History of the Twentieth-Century World* (New York: W. W. Norton, 2000), 11–13. McNeill refers to a "somatic energy regime" beginning ten thousand years ago that included human power—including slavery—and also energy derived from draft animals. In his analysis, McNeill cites the period of the industrial revolution in England as the beginning of the decline of somatic energy in Western civilization. *Natural energy* refers to water and wind power or natural energy

sources; this usage is also seen in the work of David Nye—for an example, see Nye, *Consuming Power: A Social History of American Energies* (Cambridge: MIT Press, 2001), 7.

4. Lewis Evans, *Geographical, Historical, Political, Philosophical and Mechanical Essays the First, Containing an Analysis of a General Map of the Middle British Colonies in America; and of the Country of the Confederate Indians; a Description of the Face of the Country; the Boundaries of the Confederates; and the Maritime and Inland Navigations of the Several Rivers and Lakes Contained Therein* (Philadelphia: Benjamin Franklin and D. Hall, 1755), 31. Another expression of this notion can be found in Henry Nash Smith, *Virgin Land: The American West as Symbol and Myth* (Cambridge, MA: Harvard University Press, 1950), 124.

5. Thomas Jefferson, *Notes on the State of Virginia* (London: John Stockwell, 1787), 274. Jefferson was primarily interested in the political implications of agrarianism; his view of the yeoman farmer was that of an independent social actor that fit within his conceptualization of republicanism.

6. Tench Coxe, *A View of the United States of America, in a Series of Papers, Written at Various Times, between the Years 1787 and 1794* (Philadelphia, 1794), 13.

7. *Tariff Acts Passed by the Congress of the United States from 1789 to 1895: Including All Acts, Resolutions, and Proclamations Modifying or Changing Those Acts* (Washington, DC: United States Government Printing Office, 1896), 9.

8. Thomas Jefferson et al., *The Writings of Thomas Jefferson: Containing His Autobiography, Notes on Virginia, Parliamentary Manual, Official Papers, Messages and Addresses, and Other Writings, Official and Private, Now Collected and Published in Their Entirety for the First Time* (Washington, DC: Thomas Jefferson Memorial Association, 1904), 294.

9. Jefferson, *The Writings of Thomas Jefferson: Being His Autobiography, Correspondence, Reports, Messages, Addresses, and Other Writings, Official and Private, Published by the Order of the Joint Committee of Congress on the Library, from the Original Manuscripts, Deposited in the Department of State*, vol. 5, 400.

10. Michael Williams, "Clearing the United States Forests: Pivotal Years 1810–1860," *Journal of Historical Geography* 8, no. 1 (1982): 13; see also Diane F. Britton, *The Iron and Steel Industry in the Far West: Irondale, Washington* (Niwot: University Press of Colorado, 1991), 22.

11. Alfred D. Chandler Jr., "Anthracite Coal and the Beginnings of the Industrial Revolution in the United States," *Business History Review* 46, no. 2 (1972): 144.

12. Chandler, "Anthracite Coal," 145.

13. Peter Temin, *Iron and Steel in Nineteenth-Century America, an Economic Inquiry* (Cambridge: MIT Press, 1964), 62.

14. Karl Bernard, *Travels through North America, during the Years 1825 and 1826* (Philadelphia: Carey, Lea and Carey, 1828), 158.

15. Alexander Mackay, *The Western World: Or, Travels in the United States in 1846–47: Exhibiting Them in Their Latest Development, Social, Political, and Industrial* (London: R. Bentley, 1849), 292.

16. David E. Nye, *American Technological Sublime* (Cambridge: MIT Press, 1994), 39.

17. D. Crockett, *An Account of Col. Crockett's Tour to the North and Down East: In the Year of Our Lord One Thousand Eight Hundred and Thirty-Four. His Object Being to Examine the Grand Manufacturing Establishments of the Country; and Also to Find out the Condition of Its Literature and Its Morals, the Extent of Its Commerce, and the Practical Operation of "the Experiment"* (Philadelphia, 1835), 94.

18. Charles Dickens, *American Notes for General Circulation* (London: Chapman and Hall, 1842), 193.

19. Anthony Trollope, *North America* (New York: Harper and Brothers, 1862), 247.

20. Harriet Martineau, *Society in America* (London: Saunders and Otley, 1837), 221.

21. Martineau, *Society in America*, 222.

22. Dickens, *American Notes*, 156.

23. Samuel Hopkins and Edwards Park, *The Works of Samuel Hopkins: With a Memoir of His Life and Character* (Boston: Doctrinal Tract and Book Society, 1854), 286.

24. Lawrence Jacob Friedman and Mark Douglas McGarvie, *Charity, Philanthropy, and Civility in American History* (Cambridge: Cambridge University Press, 2003), 131.

25. Ruth Clifford Engs, *Clean Living Movements: American Cycles of Health Reform* (Westport, CT: Praeger, 2000), 55–60.

26. Examples of US patents applied for and issued in the first half of the nineteenth century that specifically address smoke and spark control include the following: David Bain, Chimney cowl, US Patent X7,409, issued Feb. 5,

1833; Urban B. A. Lange, Apparatus for preventing chimneys from smoking, US Patent X8,453, issued Oct. 14, 1834; Antoine Arnoux, Gas burner (to minimize smoke), US Patent 764, issued June 4, 1838; Augustus Rice, Chimney cap, US Patent 7,275, issued Apr. 9, 1850; Jordon Mott, Improvement for chimney caps, US Patent 2,887, issued Dec. 17, 1842; Palmer Sumner, Chimney cowl, US Patent 2,964, issued Feb. 2, 1843; Stephen M. Allen, Chimney cowl, US Patent 2,568, issued Apr. 21, 1842; Joseph Hurd, Cap for regulating draft of chimneys, US Patent 3,854, issued Dec. 12, 1844; Matthias W. Baldwin, Art of managing and supplying fire for generating steam in locomotive engines, US Patent 54, issued Oct. 15, 1836; James Simpson, Construction of smoke stacks of locomotive or stationary steam engines and other chimneys for preventing the escape of sparks, US Patent 161, issued Apr. 17, 1837. Examples of US patents applied for and issued in the first half of the nineteenth century that involve innovations in nonpyrotechnic natural energy sources include: Chadiah Aylsworth, Improvement in water-wheels, US Patent 3,959, issued Mar. 21, 1845; Noahdiah W. Hubbard, Improvement in current water-wheels, being a plan for giving increased power to such wheels, US Patent 2,027, issued Apr. 2, 1841; William Zimmerman, Improvement in wind wheels, US Patent 2,107, issued May 29, 1841; Jacob Makely, Improvement in windmills, US Patent 479, issued Nov. 23, 1837; Abijah Woodward, Improvement in tub water-wheels, US Patent 1,589, issued May 8, 1840; Edward Robbins Jr. and William Ashby, Construction of water-wheels, US Patent 1,525, issued Mar. 25, 1840; F. H. Southworth, Tide or current wheel, US Patent 1,478, issued Jan. 23, 1840; and William Lewis and Thomas J. Lewis, Improvement in horizontal windmills, US Patent 583, issued Jan. 27, 1838. These patents represent a small sampling of patents issued in the first half of the nineteenth century associated with eliminating the effects of smoke and alternative energy technologies. There are several hundred in total.

27. *The Paradise within the Reach of All Men, without Labor, by Powers of Nature and Machinery: An Address to All Intelligent Men* (Pittsburgh, PA: Etzler and Reinhold, 1833), 3–4.

28. Etzler was issued two patents at the United States Patent and Trademark office, one of which shows the sophistication of his wind power technology: Navigating and propelling vessels by the action of the wind and waves, US Patent 2,533, issued Apr. 1, 1842.

29. Steven Stoll, *The Great Delusion: A Mad Inventor, Death in the Tropics, and the Utopian Origins of Economic Growth* (New York: Hill and Wang, 2008), 53.

30. William Conant Church, *The Life of John Ericsson*, vol. 2 (London: Sampson, Low, Marston, 1892), 270.

31. David E. Nye, *Consuming Power: A Social History of American Energies* (Cambridge: MIT Press, 2001), 68.

32. Ralph Waldo Emerson, "Conduct of Life," in *The Prose Works of Ralph Waldo Emerson: In Two Volumes* (Boston: J. R. Osgood, 1875), 362.

33. *Meigs County Deed Index, Pre-1830*, accessed Oct. 17, 2014, http://www.rootsweb.ancestry.com/~ohmeigs/deeds/deed_index.html?cj=1&netid=cj&o_xid=0001231185&o_lid=0001231185&o_sch=Affiliate+External.

34. W. W. Mather, *First Annual Report on the Geological Survey of the State of Ohio* (Columbus, OH: S. Medary, 1838), 150.

35. Andrew Roy, *The Coal Mines: Containing a Description of the Various Systems of Working and Ventilating Mines, Together with a Sketch of the Principal Coal Regions of the Globe, Including Statistics of the Coal Production* (Cleveland, OH: Robison, Savage, 1876), 298–335.

36. Mather, *First Annual Report*, 150.

37. Chandler, "Anthracite Coal."

38. For a more extensive breakdown and commentary of technological determinism and technological momentum, especially as it applies to systems and infrastructure, see Nye, *Consuming Power*. See also Smith and Marx, *Does Technology Drive History*.

39. Karl Marx and Frederick Engels, *Capital, Volume One: A Critique of Political Economy* (London: Dover Publications, 2012), 411–12.

40. R. H. Thurston, *A History of the Growth of the Steam-Engine* (New York: D. Appleton, 1878), 2.

41. Walter Licht, *Industrializing America: The Nineteenth Century* (Baltimore, MD: Johns Hopkins University Press, 1995), 26.

42. For more on Taqī al-Dīn's contribution to early steam engine development see Roshdi Rāshid and Régis Morelon, *Encyclopedia of the History of Arabic Science: Technology, Alchemy and Life Sciences* (London: Routledge, 1996), 779. The pump-engine, or "máquina de vapor," of Jerónimo de Ayanz y Beaumont preceded the work of Thomas Savery, who is commonly credited with its development—see R. R. A. Matas, *Bielas Y Álabes 1826–1914* (Ministerio de Industria, Turismo y Comercio. Oficina Española de Patentes y Marcas, 2008), 40. Considered by many to be the inventor of the first practical steam engine, Thomas Newcomen was the first to utilize a piston vacuum technique, in his "fire engine" of 1710, which was an improvement on Savery's design (and the subject of a patent dispute). One of the best sources

for information on the Newcomen engine design is J. Farey, *A Treatise on the Steam-Engine, Historical, Practical and Descriptive* (London: Longman, 1827), 126. For a summary of the contributions of Matthew Boulton and James Watt to the development of steam power, see William Rosen, *The Most Powerful Idea in the World: A Story of Steam, Industry, and Invention* (Chicago: University of Chicago Press, 2012), 112.

43. For a more complete history of the adoption of steam power in England see Paul Mantoux, *The Industrial Revolution in the Eighteenth Century: An Outline of the Beginnings of the Modern Factory System in England* (London: Taylor and Francis, 1928), 335. For a discussion of the number of steam engines estimated to be in operation in the United States early in the nineteenth century see Nye, *Consuming Power* (1999), Kindle ebook, location 515.

44. Elijah Hebert Luke Galloway, *History and Progress of the Steam Engine: With a Practical Investigation of Its Structure and Application* (London: T. Kelly, 1836), 298.

45. Diana Muir, *Reflections in Bullough's Pond: Economy and Ecosystem in New England* (Lebanon, NH: University Press of New England, 2000), 174.

46. J. C. Merriam, "Steam," in *Eighty Years Progress of the United States, Showing the Various Channels of Industry and Education through Which the People of the United States Have Arisen from a British Colony to Their Present National Importance* (Hartford, CT: L. Stebbins, 1865), 253. See also Paul P. Christensen, "Land Abundance and Cheap Horsepower in the Mechanization of the Antebellum United States Economy," *Explorations in Economic History* 18, no. 4 (1981): 321.

47. Nathan Rosenberg and Manuel Trajtenberg, "A General Purpose Technology at Work: The Corliss Steam Engine in the Late 19th Century U.S.," National Bureau of Economic Research Working Paper Series No. 8485 (Cambridge, MA: National Bureau of Economic Research, 2001), 38.

48. Thomas Jefferson, *Memoirs, Correspondence and Private Papers of Thomas Jefferson, Late President of the United States*, ed. Thomas Jefferson Randolph (London: Colburn and Bentley, 1829), 393. Jefferson's memoirs included correspondence regarding different types of oils used for illumination. For whale oil statistics in the United States, see *Hazard's United States Commercial and Statistical Register* (Philadelphia: W. F. Geddes, 1842), 184.

49. United States House of Representatives, *House Documents* (Washington, DC: US Government Printing Office, 1844), 329. The 1844 record of

the US House of Representatives includes a thorough discussion of lamps and their relative merits along with a brief discussion of fuels. Charles M. Keller, who was the United States Patent Office's sole patent examiner at the time, presented to the House a report on fuels and lamps. Keller's report was in response to a request by the legislature to report on "the existing condition of the arts and sciences."

50. Jacob A. Riis, *Children of the Tenements* (New York, London: Macmillan, 1904), 185. Charles S. Francis, Joseph H. Francis, and Alexander Anderson, *The Parlour Book* (Boston: Charles S. Francis, 1839), 164.

51. Williams, "Clearing the United States Forests," 15.

52. See both Marcus Bull, *Experiments to Determine the Comparative Value of the Principal Varieties of Fuel Used in the United States, and Also in Europe* (Philadelphia: J. Dobson, 1827), and Williams, "Clearing the United States Forests," 18.

53. United States Bureau of the Census, *Historical Statistics of the United States 1789–1945* (Washington, DC: Government Printing Office, 1949), 142.

54. Sam H. Schurr and Bruce Carlton Netschert, *Energy in the American Economy, 1850–1975; an Economic Study of Its History and Prospects* (Baltimore, MD: Johns Hopkins University Press, 1960), 69–70.

55. B. D. Hong and E. R. Slatick, "Carbon Dioxide Emission Factors for Coal," US Energy Information Administration, accessed Oct. 8, 2013, http://www.eia.gov/coal/production/quarterly/co2_article/co2.html. The relative figures of carbon dioxide for the years 1860 and 1890 were calculated by using the government-supplied factor of 2.8 tons of carbon dioxide for each ton of coal burned (from the web page cited above) and the coal consumption figures from the work of Schurr and Netschert, *Energy in the American Economy*, 69–70.

56. "Cincinnati," *Harper's New Monthy Magazine*, July 1883, 246; also *Leading Manufacturers and Merchants of Cincinnati and Environs: The Great Railroad Centre of the South and Southwest* (New York: International Publishing, 1886), 36.

57. Cleveland Division of Health, *Annual Report* (1881), 5; E. Cleave, "City of Cleveland and Cuyahoga County," in *Cleave's Biographical Cyclopaedia of the State of Ohio* (Cleveland, OH, 1875), 105–6.

58. *Report of Smoke Committee of the Citizens' Association of Chicago: May, 1889* (Chicago: G. E. Marshall, 1889), 38.

CHAPTER 3: THE CONUNDRUM OF SMOKE AND VISIBLE ENERGY

Epigraph: Josiah Strong, *Our Country: Its Possible Future and Its Present Crisis* (New York: Baker and Taylor, 1885), 11.

1. Frank Uekötter, *The Age of Smoke: Environmental Policy in Germany and the United States, 1880–1970* (Pittsburgh, PA: University of Pittsburgh Press, 2009), 20.

2. Angela Gugliotta, "Class, Gender, and Coal Smoke: Gender Ideology and Environmental Injustice in Pittsburgh, 1868–1914," *Environmental History* 5, no. 2 (2000): 173–74. For specific efforts against smoke in Chicago, Baltimore, Saint Louis, and Boston, see Uekötter, *Age of Smoke*, 22–25. In addition to Gugliotta and Uekötter, environmental historian David Stradling's research documents smoke-related problems in several Midwest cities; see *Smokestacks and Progressives: Environmentalists, Engineers and Air Quality in America, 1881–1951* (Baltimore, MD: Johns Hopkins University Press, 1999), 21–25.

3. Stradling, *Smokestacks and Progressives*, 2.

4. Strong, *Our Country*, 58–59, 129.

5. Strong, *Our Country*, 174.

6. Washington Gladden, *Working People and Their Employers* (New York: Funk, 1888), 15, 175.

7. An electronic content analysis of both Strong's *Our Country* and Gladden's *Working People* found no instances of the words *smoke* or *soot*.

8. Strong, *Our Country*, 131.

9. "The Public Be ———," *Chicago Journal of Commerce and Metal Industries* 61, no. 16, Oct. 30, 1892, 22.

10. "City of St. Louis v. Heitzeberg Packing & Provision Co.," *Southwestern Reporter*, vol. 42, Oct. 18, 1897–Jan. 3, 1898 (Saint Paul, MN: West Publishing, 1898), 956.

11. "City of St. Louis," 956.

12. Various courts maintained this position, a typical case being that in 1892 of *Marshall Field & Co. et al. v. City of Chicago*. See E. B. Smith and M. L. Newell, *Reports of Cases Decided in the Appellate Courts of the State of Illinois* (Chicago: Callaghan, 1893), 410.

13. *Engineering Magazine*, vol. 12, 1897, 809–11.

14. William Thomson, "The Injurious Effects of the Air of Large Towns on Animal and Vegetable Life, and Methods Proposed for Securing a Sa-

lubrious Air," *Van Nostrand's Eclectic Engineering Magazine*, vol. 20, June 1879, 491.

15. *Coal and Coal Trade Journal* 42 (1911): 400.

16. David V. Mollenhoff, *Madison, a History of the Formative Years* (Madison: University of Wisconsin Press, 2003), 124.

17. Samuel Hazard, *The Register of Pennsylvania: Devoted to the Preservation of Facts and Documents and Every Other Kind of Useful Information Respecting the State of Pennsylvania* (Pittsburgh, PA: W. F. Geddes, 1829), 31.

18. Booth Tarkington, *The Turmoil: A Novel* (New York: Harper and Brothers, 1915), 1–6.

19. *Fuel Magazine: The Coal Operators National Weekly*, 1909, 21.

20. "Operator's Face Slapped," *New York Times*, Sept. 15, 1897.

21. E. E. Carreras, "St. Louis Siftings," *Weekly Northwestern Miller*, June 8, 1888, 579.

22. "The Smoke Nuisance and Its Regulation, with Special Reference to the Condition Prevailing in Philadelphia," *Journal of the Franklin Institute* 144, July 1897, 31.

23. *Smoke Investigations: Bulletin, No. 1–10: 1912–1922* (Pittsburgh, PA: University of Pittsburgh, Mellon Institute of Industrial Research, 1922), 85–86.

24. Eduard von Grauvogl and Geo E. Shipman, *Text Book of Homoeopathy* (Chicago: C. S. Halsey, 1870), 169. See also "Employment of Mine Labor," *Mining and Metallurgy* 148, Apr. 1919, 673.

25. Donald MacMillan, *Smoke Wars: Anaconda Copper, Montana Air Pollution, and the Courts, 1890–1924* (Helena: Montana Historical Society Press, 2000), 27.

26. For a synopsis of when various reformers began to organize for smoke abatement see Uekötter, *Age of Smoke*, Kindle ebook, location 336.

27. John H. Griscom, *The Uses and Abuses of Air: Showing Its Influence in Sustaining Life, and Producing Disease; with Remarks on the Ventilation of Houses, and the Best Methods of Securing a Pure and Wholesome Atmosphere inside of Dwellings, Churches, Courtrooms, Workshops, and Buildings of All Kinds* (New York: Redfield, 1848), 163, 76, 226, and 99.

28. "The Sewage Question: The Relation of Town and Country," *Farmers Magazine*, Nov. 1859, 379.

29. James Copeland, *A Dictionary of Practical Medicine: Comprising General Pathology*, vol. 1 (New York: Harper and Brothers, 1845), 186.

30. James Copeland, *A Dictionary of Practical Medicine*, vol. 3 (London: Longmans, Brown, Green, Longmans and Roberts, 1858), 421.

31. Massachusetts Medical Society, *Medical Communications*, vol. 8 (Boston: Massachusetts Medical Society, 1854), 107.

32. *American Journal of the Medical Sciences*, vol. 18 (Philadelphia: Lea and Blanchard), 132.

33. Stephen Smith, "Our Sanitary Condition," *New York Times*, 1865.

34. "The Lake Tragedy," *Chicago Tribune*, Feb. 17, 1874.

35. Deaths due to indoor air quality and asphyxiation caused by smoke or coal gas were not uncommon, with historical accounts indicating individuals from a variety of backgrounds as victims. See "Two Young Ladies Meet with an Untimely End," *Cleveland Plain Dealer*, Jan. 13, 1864; "A Fatal Sleep," *Cincinnati Enquirer*, Dec. 18, 1886; "What We Foul Atmosphere of Badly Ventilated Rooms," *Cincinnati Enquirer*, Jan. 30, 1869; "Suffocation by Coal Gas," *Chicago Tribune*, Jan. 4, 1866; "Killed by Coal Gas," *Chicago Tribune*, Dec. 23, 1877; "A Whole Family . . . by Coal Gas," *Chicago Tribune*, Dec. 28, 1868.

36. Jill Jonnes, *Empires of Light: Edison, Tesla, Westinghouse, and the Race to Electrify the World* (New York: Random House, 2003), 137.

37. David E. Nye, *America as Second Creation: Technology and Narratives of New Beginnings* (Cambridge: MIT Press, 2003), 192.

38. Charles Frederick Chandler, *Report on the Gas Nuisance in New York* (New York: D. Appleton, 1870), 42.

39. J. A. Mathews, *Report upon Smoke Abatement: An Impartial Investigation of the Ways and Means of Abating Smoke, Results Attained in Other Cities, Merits of Patented Devices, Together with Practical Suggestions to the Department of Smoke Abatement, the Steam Plant Owner and the Private Citizen* (Syracuse, NY: Syracuse Chamber of Commerce, 1907), 10.

40. Gugliotta, "Class, Gender, and Coal Smoke," 166–67.

41. Catharine Esther Beecher and Harriet Beecher Stowe, *The American Woman's Home: Or, Principles of Domestic Science; Being a Guide to the Formation and Maintenance of Economical, Healthful, Beautiful, and Christian Homes* (New York: J. B. Ford, 1869), 424.

42. "How to Make Pleasant Homes," *Godey's Magazine*, vol. 92, Mar. 1876, 281.

43. George A. Gonzalez, *The Politics of Air Pollution: Urban Growth, Ecological Modernization, and Symbolic Inclusion* (Albany: State University of New York Press, 2005), 47.

44. Stradling, *Smokestacks and Progressives*, 44–45.

45. John S. Reese, *Guide Book for the Tourist and Traveler over the Valley Railway: The Short Line between Cleveland, Akron, and Canton* (Kent, OH: Kent State University Press in cooperation with Cuyahoga Valley National Park and Cuyahoga Valley Scenic Railroad, 2002), 40.

46. "Ventilation," *Cleveland Medical Gazette* 1, no. 7, Jan. 1, 1860, 191.

47. "Ventilation," 191.

48. Edward Peron Hingston, *The Genial Showman. Being Reminiscences of the Life of Artemus Ward; and Pictures of a Showman's Career in the Western World* (London: J. C. Hotten, 1871), 86.

49. David Stradling and Peter Thorsheim, "The Smoke of Great Cities: British and American Efforts to Control Air Pollution, 1860–1914," *Environmental History* 4, no. 1 (1999): 13.

50. W. Goss, *Smoke Abatement and Electrification of Railway Terminals in Chicago*, Report of the Chicago Association of Commerce, Committee of Investigation on Smoke Abatement and Electrification of Railway Terminals (Chicago: Chicago Association of Commerce, Committee of Investigation on Smoke Abatement Industry, 1915), 82.

51. "Public Be ———," 22.

52. Chicago Department of Health, *Report of the Dept. of Health of the City of Chicago* (1891), 128.

53. Christine Meisner Rosen, "Businessmen against Pollution in Late Nineteenth Century Chicago," *Business History Review* 69, no. 3 (1995): 352. Rosen's article is an in-depth case study of smoke abatement efforts in Chicago after 1890. Another source that describes Chicago's air quality in the era is William Cronon, *Nature's Metropolis: Chicago and the Great West* (New York: W. W. Norton, 1992).

54. Rosen, "Businessmen against Pollution," 356.

55. Martin V. Melosi, *Effluent America: Cities, Industry, Energy, and the Environment* (Pittsburgh, PA: University of Pittsburgh Press, 2001). For a comprehensive review of the literature regarding smoke abatement efforts, see Melosi's summary, 18–20.

56. Edward Bryant, *Climate Process and Change* (Cambridge: Cambridge University Press, 1997), 121. Bryant provides a concise history of the early discoveries linking carbon dioxide from fossil fuels to climate change both through the work of Tyndall, Arrhenius, and later Thomas Chamberlain. According to Mark Maslin, *Global Warming: A Very Short Introduction* (Oxford: Oxford University Press, 2004), it was not until 1987 that the

pivotal role of carbon dioxide in climate change was confirmed, through an analysis of ice core samples from the Antarctic Vostok experiments.

57. Rosen, "Businessmen against Pollution," 352.

58. Cutler J. Cleveland, *Concise Encyclopedia of the History of Energy* (San Diego, CA: Elsevier, 2009), 165.

59. See Julius Austin, Wind Wheel, US Patent 231,253, issued Aug. 17, 1880; J. C. Preston, Wind Wheel, US Patent 232,205, issued Sept. 14, 1880; H. B. Smith, Wind Wheel, US Patent 232,558, issued Sept. 21, 1880; Otis D. Thompson, Wind Wheel, US Patent 235,470, issued Dec. 14, 1880; T. B. Gray, Wind Wheel, US Patent 232,815, issued Oct. 5, 1880; George A. Myers, Wind Wheel, US Patent 229,907, issued July 13, 1880; Jacob Gilstrap, Wind Wheel, US Patent 236,018, issued Dec. 28, 1880; Ransom E. Strait, Wind Engine, US Patent 225,539, issued Mar. 16, 1880; Henry Croft Sr. and Henry Croft Jr., Wind Wheel, US Patent 224,817, issued Feb. 24, 1880; John Robert Brink, Wind Engine, US Patent 232,673, issued Sept. 28, 1880; Benjamin M. Rolph, Windmill, US Patent 234,204, issued Nov. 9, 1880; John F. Garatt, Windmill, US Patent 234,975, issued Nov. 30, 1880.

60. See Adolphus C. Kendel, Smoke-bell, US Patent 231,587, issued Aug. 24, 1880; Asa W. Lafrance, Smoke-stack, US Patent 227,272, issued May 4, 1880; David Sinton, Smoke-consuming furnace, US Patent 233,168, issued Oct. 12, 1880; William E. Johnson, Fire-safe for smoke-houses, US Patent 226,861, issued Apr. 27, 1880; Charles T. Barnes, Smoke and gas consumer for stoves, US Patent 231,142, issued Aug. 17, 1880; Arthur Staples, Smoke-stack, US Patent 230,673, issued Aug. 3, 1880; William C. P. Bissell, Smoke and gas consuming furnace, US Patent 234,385, issued Nov. 16, 1880; George S. Roberts, Smoke-bell support for gas fixtures, US Patent 230,060, issued July 13, 1880; Huntington Brown, Smoke-stack, US Patent 226,439, issued Apr. 13, 1880; David Grosebeck, Spark-arrester, US Patent 235,762, issued Dec. 21, 1880; John E. Sampsel, Smoke-box and stack for locomotive, US Patent 227,657, issued May 18, 1880; Patrick Quinn, Upright steam-boiler, US Patent 226,880, issued Apr. 27, 1880; Anton Pohl, Spark-arrester, US Patent 230,568, issued July 27, 1880; Pratt Wright, Spark-arrester for locomotives, US Patent 228,431, issued June 1, 1880; George A. Gunther, Locomotive spark-extinguisher, US Patent 234,274, issued Nov. 9, 1880; David J. Timlin, Spark-arrester, US Patent 234,349, issued Nov. 9, 1880; Peter A. Aikman, Stove damper, US Patent 233,456, issued Oct. 19, 1880; James M. Russell, Spark-arrester, US Patent 229,642, issued July 6, 1880; William Gray, Furnace, US Patent 233,614, issued Oct. 26, 1880; Chauncey Lamou-

reux, Chimney cap, US Patent 230,483, issued July 27, 1880; Rufus S. Craig, Spark-arrester, US Patent 233,400, issued Oct. 19, 1880; Karl W. Neuhaus, Spark-arrester, US Patent 228,922, issued June 15, 1880; Elias H. McNiel, Steam-boiler and furnace, US Patent 225,625, issued Mar. 16, 1880; Hampton E. Campfield, Spark-arrester, US Patent 228,442, issued June 8, 1880; Oliver Bryan, Hot-air furnace, US Patent 234,232, issued Nov. 9, 1880; Benjamin W. Felton, Ventilator for chimney-caps, US Patent 226,507, issued Apr. 13, 1880; George L. Blewett, Spark-extinguisher, US Patent 225,326, issued Mar. 9, 1880, John E. Wiggin, Spark-arrester, US Patent 226,378, issued Apr. 6, 1880; Henry A. Ridley, Spark-arrester, US Patent 233,022, issued Oct. 5, 1880; John P. Putnam, Ventilating-gasolier, US Patent 233,372, issued Oct. 19, 1880; George B. F. Cooper, Spark-arrester, US Patent 225,572, issued Mar. 16, 1880; Edward Dunn, Spark-arrester, US Patent 227,420, issued May 11, 1880; James Suitt, Spark-arrester, US Patent 224,802, issued Feb. 24, 1880; Michael Zeck, Spark-arrester, US Patent 223,427, issued Jan. 6, 1880; John E. Wiggin, Spark-arrester, US Patent 224,497, issued Feb. 10, 1880; Leonard Henderson, Smoke and dust arrester for railway cars, US Patent 225,448, issued Mar. 9, 1880; William M. Thorton, Spark-arrester, US Patent 229,207, issued June 22, 1880; Alexander Rideout, Warming and ventilating, US Patent 229,842, issued July 13, 1880; Francis A. Perky, Spark-arrester, US Patent 153,907, reissued Aug. 31, 1880; Charles C. Caywood, Chimney cap and cowl, US Patent 224,873, Feb. 24, 1880; N. N. Horton, Heater, cooler, and ventilator for railroad cars, US Patent 227,977, issued May 25, 1880; John S. Lloyd, Smoke and cinder conveyor for locomotives, US Patent 227,550, issued May 11, 1880; John W. Mark, Ventilator, US Patent 230,952, issued Aug. 10, 1880; Isaac Bales, Combined house ventilator and register, US Patent 232,166, issued Sept. 14, 1880; Hans J. Andersen, Heater and ventilator, US Patent 235,486, issued Dec. 14, 1880; James F. Baldwin, Ventilator for dwellings, US Patent 233,962, issued Nov. 2, 1880; Harry Pursell, Grate-front, US Patent 225,074, issued Mar. 2, 1880; J. Nelson Harris, Grate and fender for fire-places, US Patent 228,994, issued June 2, 1880; Alexander Mitchell, Cinder-guard, US Patent 226,869, issued Apr. 27, 1880; Andrew T. Jackson, Chimney flue and shield, US Patent 226,074, issued Mar. 30, 1880.

    61. "Smoke Prevention: Report of the Special Committee on Prevention of Smoke," *Journal of the Association of Engineering Societies* 11 (1892): 321. These figures are based on the delivered price at Chicago in 1892.

    62. George R. Ide, "Smoke Abatement in Cities," *Proceedings of the Engineers' Club of Philadelphia* 9, no. 2 (1892): 151.

63. Citizens' Association of Chicago, Committee on Smoke, *Report of Smoke Committee of the Citizens' Association of Chicago: May, 1889* (Chicago: G. E. Marshall, 1889), 6.

64. "The Prevention of Smoke," *American Gas Light Journal and Chemical Repertory*, June 2, 1880, 249.

65. "Washing Smoke," *Scribner's Monthly*, Jan. 1876, 453.

66. For a good example of this technique see *New York Farmer and American Gardener's Magazine*, vol. 8, 1835, 288.

67. American Railway Master Mechanics' Association, *Annual Report of the American Railway Master Mechanics' Association* (Cincinnati, OH: Aldens, 1880), 89–90.

68. New York Department of Health and New York Board of Health, *Annual Report of the Board of Health of the Department of Health of the City of New York*, vol. 2 (1872), 71–72.

69. John M. Squire, John P. Burpee, and George C. Robson, *The Official Record of the State Board of Health of Massachusetts Together with a Phonographic Report of the Evidence and Arguments at the Hearing* (Cambridge, MA: Welch, Bigelow, 1874), 381.

70. Robert M. Bancroft and Francis J. Bancroft, *Tall Chimney Construction. A Practical Treatise on the Construction of Tall Chimney Shafts Constructed in Brick, Stone, Iron and Concrete* (Manchester: J. Calvert, 1885), 9.

71. American Society of Mechanical Engineers, *Transactions of the American Society of Mechanical Engineers* (New York: American Society of Mechanical Engineers, 1894), 1178.

72. See, for example, George Richardson Porter, *Useful Arts: A Treatise on the Origin, Progressive Improvement, and Present State of the Silk Manufacture* (Philadelphia: Carey and Lea, 1832), 155. See also Nathan Hale, *Chronicle of Events, Discoveries, and Improvements, for the Popular Diffusion of Useful Knowledge, with an Authentic Record of Facts. Illustrated with Maps and Drawings* (Boston: S. N. Dickinson, 1850), 509.

73. One of the best works dealing with the emergence of engineering as it relates to nineteenth- and twentieth-century technology is David F. Noble, *America by Design: Science, Technology, and the Rise of Corporate Capitalism* (Oxford: Oxford University Press, 1979). Although Noble's main argument is that the rise in technology is synonymous with the rise of corporate capitalism, the work also goes far in explaining the interplay

between technology and society within the historical construct of technological determinism. A good primer is found in the book's foreword by Christopher Lasch on page xi.

74. United States Bureau of the Census, *Historical Statistics of the United States, Colonial Times to 1970* (Washington, DC: US Department of Commerce, Bureau of the Census, 1975). The rising output in consumer goods in the United States is closely correlated with energy consumption. For a breakdown of total US output of semidurable and perishable goods from 1869 forward, for example, see page 699. For total horsepower of all prime movers beginning in 1849, see page 818.

75. The position that modern environmentalism was more of an outgrowth of concerns for quality of life than an inherent concern for the environment was first posited by Samuel P. Hays and Barbara D. Hays, in *Beauty, Health, and Permanence: Environmental Politics in the United States, 1955–1985* (Cambridge: Cambridge University Press, 1987), 5.

CHAPTER 4: TECHNOLOGY AND ENERGY IN THE ABSTRACT

Epigraph: Abraham Lincoln, *Speeches and Writings Part 2: 1859–1865: Library of America 46* (New York: Library of America, 1989), 10.

1. Christopher F. Jones, *Routes of Power: Energy and Modern America* (Cambridge, MA: Harvard University Press, 2014), 2.

2. Typical of the era were articles such as "Annihilation of Space," *Brooklyn Daily Eagle*, Nov. 29, 1844, 2. See also "The Completion of the Pacific Railroad," *Charleston Daily News*, May 12, 1869, 2.

3. *One Hundred Years' Progress of the United States* (Hartford, CT: L. Stebbins, 1870), v.

4. Lincoln, *Speeches and Writings*, 8.

5. Walt Whitman and G. Schmidgall, *Walt Whitman: Selected Poems 1855–1892* (New York: St. Martin's, 2000), 3.

6. Ralph Waldo Emerson, *The Collected Works of Ralph Waldo Emerson: Society and Solitude*, ed. R. E. Spiller, A. R. Ferguson, J. Slater, and J. F. Carr (Cambridge, MA: Belknap Press of Harvard University Press, 1971), 80.

7. Emerson, *Collected Works*, 80.

8. Leonard N. Neufeldt, "The Science of Power: Emerson's Views on Science and Technology in America," *Journal of the History of Ideas* 38, no. 2 (1977): 329–44.

9. Jacob Bigelow, *Elements of Technology: Taken Chiefly from a Course of Lectures Delivered at Cambridge, on the Application of the Sciences to the Useful Arts* (Boston: Hilliard, Gray, Little and Wilkins, 1829), 4.

10. Merritt Roe Smith and Leo Marx, eds., *Does Technology Drive History? The Dilemma of Technological Determinism* (Cambridge: MIT Press, 1994), 5.

11. Perry Miller, "The Responsibility of Mind in a Civilization of Machines," *American Scholar* 31, no. 1 (1961): 53–55.

12. Leo Marx, *The Machine in the Garden: Technology and the Pastoral Ideal in America* (New York: Oxford University Press, 1964), 121, 197.

13. Christopher Lasch, *The Minimal Self: Psychic Survival in Troubled Times* (New York: W. W. Norton, 1984), 43. Another scholar who explores trends of confidence in technology is David Noble. In *Religion of Technology*, Noble identifies Edison and the age of electricity in terms of a technological utopia. See chapter 5 in *The Religion of Technology: The Divinity of Man and the Spirit of Invention* (New York: A. A. Knopf, 1997).

14. David E. Nye, *American Technological Sublime* (Cambridge: MIT Press, 1994), xiii.

15. See Richard F. Hirsh and Benjamin K. Sovacool, "Wind Turbines and Invisible Technology: Unarticulated Reasons for Local Opposition to Wind Energy," *Technology and Culture* 54, no. 4 (2013): 720. Hirsh and Sovacool here are citing Borg, "Les sens perdus du garagiste: Comment le savoir-faire a été disqualifié dans l'univers automobile américain," *Revue d'histoire moderne et contemporaine* 59, no. 3 (2012): 22.

16. William Cronon, *Nature's Metropolis: Chicago and the Great West* (New York: W. W. Norton, 1992), 236.

17. Robert W. Rydell, *All the World's a Fair: Visions of Empire at American International Expositions, 1876–1916* (Chicago: University of Chicago Press, 2013), 9; Sean Dennis Cashman, *America in the Gilded Age: From the Death of Lincoln to the Rise of Theodore Roosevelt* (New York: New York University Press, 1984), 8.

18. "Steam Street Cars," *Times Philadelphia*, Aug. 21, 1876, 4. See also "Terrible Boiler Explosion," *New York Times*, Nov. 12, 1867, 4.

19. Frank Dalzell, *Engineering Invention: Frank J. Sprague and the U.S. Electrical Industry* (Cambridge: MIT Press, 2010), 33.

20. "Characteristics of the International Fair," *Atlantic Monthly*, Sept. 1876, 359.

21. Rydell, *All the World's a Fair*, 16.

22. Rydell, *All the World's a Fair*, 15.

23. Rydell, *All the World's a Fair*, 867.

24. Barbara Freese provides a comprehensive account of the labor issues between Gowen and the miners; see Freese, *Coal: A Human History* (Cambridge, MA: Perseus, 2003), 130.

25. Douglas O. Linder, "The Molly Maguires Trials: A Chronology," *Famous Trials: The Molly Maguires Trials (1876–77)*, 2010, accessed Aug. 10, 2014, http://law2.umkc.edu/faculty/projects/ftrials/maguires/mollieschrono.html, confirms that anthracite coal was used for the boilers on the grounds of the Centennial Exhibition.

26. United States Centennial Commission, *International Exhibition, 1876*, ed. A. T. Goshorn, Francis Walker, and Dorsey Gardner (Washington, DC: US Government Printing Office, 1880), 194. For water consumption and more in-depth specifications see H. F. Mueller, "The 'One Hundredth Anniversary' of George Henry Corliss," *Power*, May 14, 1918, 686.

27. Pennsylvania Board of Centennial Managers, *Pennsylvania and the Centennial Exposition: Comprising the Preliminary and Final Reports of the Pennsylvania Board of Centennial Managers, Made to the Legislature at the Sessions of 1877–8* (Philadelphia: Gillin and Nagle, 1878), 177.

28. J. M. Wilson, "The Centennial International Exhibition of 1876—No. XIV," *Engineering*, Apr. 28, 1876, 331.

29. "City Noises," *New York Times*, June 6, 1879.

30. George Basalla, "Some Persistent Energy Myths," in *Energy and Transport: Historical Perspectives on Policy Issues*, ed. George H. Daniels and Mark H. Rose (Beverly Hills: Sage, 1982), 28.

31. General Assembly Presbyterian Church in the USA, *Centennial Historical Discourses: Delivered in the City of Philadelphia, June, 1876* (Philadelphia: Presbyterian Board of Publication, 1876), 242–43.

32. Rev. T. M. Post, "The Outlook of the Times in Reference to the Progress of Christianity," *Missionary Herald* 78, no. 1, Jan. 1882, 20..

33. W. J. Hammer, "William Wallace, and His Contributions to the Electrical Industries—II," *Electrical Engineer* (Feb. 8, 1893): 129.

34. Robert V. Bruce, *Bell: Alexander Graham Bell and the Conquest of Solitude* (Ithaca, NY: Cornell University Press, 1990), 194, 197.

35. Silvanus P. Thompson, *The Life of Lord Kelvin* (Providence, RI: American Mathematical Society, 2005; first published 1976 by Chelsea Publishing), 671.

36. Alfred Russell Wallace, *The Progress of the Century* (New York: Harper and Brothers, 1901), 276.

37. United States Centennial Commission, *Reports and Awards* (Philadelphia: J. B. Lippincott, 1877), 15.

38. Elihu Thomson, "Electricity in 1876 and in 1893," *Electrical Review*, Jan. 26, 1894, 104.

39. Thomson, "Electricity in 1876," 104.

40. Thomas A. Edison to William Wallace, Sept. 13, 1878, D7819G; TAEM 17:925, Thomas A. Edison Papers Digital Edition, Rutgers University.

41. *Webster's Unabridged Dictionary, Comprising the Issues of 1864, 1879, and 1884*, ed. Noah Porter (Springfield, MA: Merriam, 1898), s.v. "electricity."

42. Independence Hall Pennsylvania Academy of the Fine Arts, *Proceedings of the One Hundredth Anniversary of the Introduction and Adoption of the "Resolutions Respecting Independency." Held in Philadelphia on the Evening of June 7, 1876, at the Pennsylvania Academy of the Fine Arts, and on July 1, 1876, at the Hall of Independence* (Philadelphia, 1876), 64.

43. Independence Hall Pennsylvania Academy of the Fine Arts, *Proceedings*, 14.

44. Independence Hall Pennsylvania Academy of the Fine Arts, *Proceedings*, 17.

45. Independence Hall Pennsylvania Academy of the Fine Arts, *Proceedings*, 82.

46. "Centennial Notes. No. 15. England in Memorial Hall," *Friends' Intelligencer*, Sept. 1876, 457.

47. "Centennial Notes," 477.

48. Post, "Outlook of the Times," 20.

49. Mrs. M. B. Norton, "A Centennial Outlook," in *Life and Light for Heathen Women*, vol. 6 (Boston: Rand, Avery, 1876), 109, 11–12.

50. Olav Thulesius, *The Man Who Made the Monitor: A Biography of John Ericsson, Naval Engineer* (Jefferson, NC: McFarland, 2007), 164.

51. "The Subscriptions to the Centennial $4,150,000," *New York Times*, Mar. 31, 1874.

52. John Ericsson, *Contributions to the Centennial Exhibition* (New York, 1876), iii.

53. Ericsson, *Contributions*, 575–77.

54. *Annual of Scientific Discovery, 1869* (Boston: Gould and Lincoln, 1869), 65.

55. George Basalla, *The Evolution of Technology* (Cambridge: Cambridge University Press, 1988), 25.

56. David E. Nye, *Electrifying America: Social Meanings of a New Technology, 1880–1940* (Cambridge: MIT Press, 1990), 37.

CHAPTER 5: OF FLUIDS, FIELDS, AND WIZARDS

Epigraph: "Electrical Magic," *Scientific American,* 45, no. 21 (Nov. 19, 1881): 327.

1. Michael B. Schiffer, Kacy L. Hollenback, and Carrie L. Bell, *Draw the Lightning Down: Benjamin Franklin and Electrical Technology in the Age of Enlightenment* (Berkeley: University of California Press, 2003), 1.

2. I. Bernard Cohen, *Benjamin Franklin's Science* (Cambridge, MA: Harvard University Press, 1990), 66–67.

3. Preface to *Experiments and Observations on Electricity, Made at Philadelphia in America*, by Benjamin Franklin (London, 1769), iv.

4. For a discussion on "theory-constitutive metaphors" and how they are utilized to explain and envision complex scientific theories see Eric Charles Steinhart, *The Logic of Metaphor: Analogous Parts of Possible Worlds* (Dordrecht, Netherlands: Kluwer Academic Publishers, 2001), 8.

5. Cohen, *Benjamin Franklin's Science*, 10.

6. Richard P. Olenick, Tom M. Apostol, and David L. Goodstein, *Beyond the Mechanical Universe: From Electricity to Modern Physics* (Cambridge: Cambridge University Press, 1986), 128. John L. Heilbron, *Electricity in the 17th and 18th Centuries: A Study of Early Modern Physics* (Berkeley: University of California Press, 1979), 6.

7. Marcello Pera and Jonathan Mandelbaum, *The Ambiguous Frog: The Galvani-Volta Controversy on Animal Electricity* (Princeton, NJ: Princeton University Press, 1992), xii. See also Alan J. McComas, *Galvani's Spark: The Story of the Nerve Impulse* (Oxford: Oxford University Press, 2011), 17–19.

8. Heilbron, *Electricity in the 17th and 18th Centuries*, 492.

9. Heilbron, *Electricity in the 17th and 18th Centuries*, 494.

10. Iwan Rhys Morus, *When Physics Became King* (Chicago: University of Chicago Press, 2005), 94.

11. Richard P. Olenick, Tom M. Apostol, and David L. Goodstein, *The Mechanical Universe: Introduction to Mechanics and Heat* (Cambridge: Cambridge University Press, 2008), 219.

12. Timothy Shanahan, "Kant, Naturphilosophie, and Oersted's Discovery of Electromagnetism: A Reassessment," *Studies in History and Philosophy of Science* 20, no. 3 (1989): 303.

13. Peter Michael Harman, *Energy, Force, and Matter: The Conceptual Development of Nineteenth-Century Physics* (Cambridge: Cambridge University Press, 1995), 75.

14. Heilbron, *Electricity in the 17th and 18th Centuries*, 494. Helibron contends that Ørsted's discovery of electromagnetism did nothing more than to "transform our civilization."

15. Olenick, Apostol, and Goodstein, *Beyond the Mechanical Universe*, 270.

16. Andrew Zangwill, *Modern Electrodynamics* (New York: Cambridge University Press, 2013), 466.

17. Quoted in John L. Heilbron, *The Oxford Companion to the History of Modern Science* (New York: Oxford University Press, 2003), 301.

18. Steinhart, *Logic of Metaphor*. Steinhart's work traces how theories have been conceptualized in language over time. In addition to electromagnetic forces, the work examines the linguistic analogies utilized for light, heat, and sound—see pages 26 and 173 for summaries on the concept of ether or aluminiferous ether.

19. Tian Yu Cao, *Conceptual Developments of 20th Century Field Theories* (Cambridge: Cambridge University Press, 1997), 28.

20. See Albert E. Moyer, *Joseph Henry: The Rise of an American Scientist* (Washington, DC: Smithsonian Institution Press, 1997), 78–111. See also Thomas Coulson, *Joseph Henry: His Life and Work* (Princeton, NJ: Princeton University Press, 1950), 65–95.

21. Joseph Henry, "On a Reciprocating Motion Produced by Magnetic Attraction and Repulsion," *American Journal of Science* 20, no. 2 (July 1831): 340.

22. Joseph Henry, *Scientific Writings of Joseph Henry* (Washington, DC: Smithsonian Institution, 1886), 434.

23. Mary A. Henry, "The Electro-Magnet; or Joseph Henry's Place in the History of the Electro Magnetic Telegraph," *Electrical Engineer* 17, no. 297 (Jan. 10, 1894): 26.

24. Coulson, *Joseph Henry*, 63.

25. See Coulson, *Joseph Henry*, 63, and Moyer, *Joseph Henry*; E. N. Dickerson, *Joseph Henry and the Magnetic Telegraph* (New York: C. Scribner's Sons, 1885), 46, and David Hochfelder, *The Telegraph in America, 1832–1920* (Baltimore, MD: Johns Hopkins University Press, 2012), 231. See also Maury Klein, *The Power Makers: Steam, Electricity, and the Men Who Invented Modern America*, 1st US ed. (New York: Macmillan, 2008), 109.

26. Joseph Henry, *A Memorial of Joseph Henry* (Washington, DC: US Government Printing Office, 1880), 303.

27. Daniel Sperling and Deborah Gordon, *Two Billion Cars: Driving toward Sustainability*, foreword by Gov. Arnold Schwarzenegger (New York: Oxford University Press, 2009), 101; and David Pimentel, "Ethanol Fuels: Energy Balance, Economics, and Environmental Impacts Are Negative," *Natural Resources Research* 12, no. 2 (2003): 127.

28. Moyer, *Joseph Henry*, 72.

29. Quoted in Coulson, *Joseph Henry*, 338.

30. Samuel F. B. Morse, Improvement in the Mode of Communicating Information by Signals by the Application of Electro-Magnetism, US Patent 1,647, issued June 20, 1840.

31. William Fothergill Cooke, *The Electric Telegraph: Was It Invented by Professor Wheatstone?* (London, 1854), 3.

32. William B. Taylor, *A Memoir of Joseph Henry: A Sketch of His Scientific Work* (Philadelphia: Collins, 1879), 18.

33. Michael B. Schiffer, *Power Struggles: Scientific Authority and the Creation of Practical Electricity before Edison* (Cambridge: MIT Press, 2008), 138.

34. Thomas L. Hankins, *Science and the Enlightenment* (Cambridge: Cambridge University Press, 1985), 66–67.

35. Paul L. Rossiter, *The Electrical Resistivity of Metals and Alloys* (New York: Cambridge University Press, 1987), 187.

36. Vladimir Gurevich, *Electric Relays: Principles and Applications* (Boca Raton, FL: Taylor and Francis, 2005), 13.

37. Samuel I. Prime, *The Life of Samuel F. B. Morse, LL. D.: Inventor of the Electro-Magnetic Recording Telegraph* (New York: D. Appleton, 1875), 295.

38. Edward L. Morse, ed., *Samuel F. B. Morse: His Letters and Journals* (Boston: Houghton Mifflin, 1914), 222.

39. "The Magnetic Telegraph," *Baltimore Sun*, May 31, 1844.

40. Thomas Edison's first hundred patents had to do with telegraphy, including but not limited to printing, duplexing, and quadplexing messaging.

41. Richard R. John, *Network Nation: Inventing American Telecommunications* (Cambridge, MA: Belknap Press of Harvard University Press, 2010), 200.

42. Regular press reports in anticipation of electric light were common in the 1880s. See "A New Brilliant Light: Gas Out," *Chicago Tribune*, Mar. 10, 1877; "Light by Lightening," *Chicago Tribune*, Feb. 11, 1877; "Electric Light,"

Boston Globe, July 16, 1878; "The First Electric Light," *Hartford Daily Courant*, Feb. 6, 1879; and "Driving Away Darkness," *Hartford Daily Courant*, July 12, 1879.

43. *Electrician* 7 (May 21–Nov. 12, 1881): 67.

44. See David E. Nye, *Electrifying America: Social Meanings of a New Technology, 1880–1940* (Cambridge: MIT Press, 1990), 1–28. Nye's opening chapter, "Middletown Lights Up," follows the lighting of Muncie, Indiana, as well as a number of other small towns. Newspaper articles that reported on lighting installations such as that of Wooster, Ohio, are typical; see "Ohio News," *Cleveland Plain Dealer*, Feb. 11, 1880.

45. William J. Hausman, Peter Hertner, and Mira Wilkins, *Global Electrification: Multinational Enterprise and International Finance in the History of Light and Power, 1878–2007* (New York: Cambridge University Press, 2008), 10.

46. W. Bernard Carlson, *Innovation as a Social Process: Elihu Thomson and the Rise of General Electric* (Cambridge: Cambridge University Press, 2003), 134.

47. From C. W. Weesner, *History of Wabash County, Indiana* (Chicago: Lewis Publishing, 1914), 317.

48. Weesner, *History of Wabash County*, 313.

49. Linda L. Robertson, *Wabash County History Bicentennial Edition 1976, Wabash, Indiana* (Marceline, MO: Walsworth, 1976), 478.

50. "The Brush Electric Light," *Scientific American* 44, no. 14 (Apr. 2, 1881): 211.

51. "The Rival Lights," *Cleveland Plain Dealer*, Dec. 22, 1880.

52. "Edison's Newest Marvel," *New York Sun*, Sept. 16, 1878, clipping, document 1439, Thomas A. Edison Papers Digital Edition, Rutgers University.

53. Carlson, *Innovation as a Social Process*, 134.

54. "Edison's Newest Marvel."

55. "Edison's Newest Marvel."

56. "Candles Are Dangerous," *Morning Star and Catholic Messenger* (New Orleans), Sept. 29, 1878; "Fatal Coal Oil Accident," *National Republican* (Washington, DC), Sept. 13, 1878.

57. *Pamphlets on Insurance* (New York: J. H. and C. M. Goodsell, 1871), 70–75; Sara E. Wermiel, "Did the Fire Insurance Industry Help Reduce Urban Fires in the United States in the Nineteenth Century?," in *Flammable Cities: Urban Conflagration and the Making of the Modern World*, ed. Greg

Bankoff, Uwe Lübken, Jordan Sand, and Stephen J. Pyne (Madison: University of Wisconsin Press, 2012), 245.

58. The nickname "Edison, the Magician," was coined in an article from the *Cincinnati Commercial* in 1878; see "Edison, the Magician," *Cincinnati Commercial*, Apr. 1, 1878, clipping (newspaper), document SB031094a, Thomas A. Edison Papers Digital Edition, Rutgers University.

59. John Pierpont Morgan to Calvin Goddard, Sept. 12, 1881, document X123BAF; TAEM 0:0, Thomas A. Edison Papers Digital Edition, Rutgers University; see also "Edison and the Skeptics," *New York Times*, Jan. 4, 1880.

60. For more about the crash and gradual recovery of gas stocks see "Edison's Electric Light," *Chicago Tribune*, Dec. 1, 1878. For Edison's patent see T. A. Edison, Electric lamp, US Patent 223,898, issued Jan. 27, 1880.

61. S. Van Dulken, *Inventing the 19th Century: 100 Inventions That Shaped the Victorian Age from Aspirin to the Zeppelin* (New York: New York University Press, 2001), 80.

62. Morgan to Goddard, Sept. 12, 1881.

63. Jill Jonnes, *Empires of Light: Edison, Tesla, Westinghouse, and the Race to Electrify the World* (New York: Random House, 2003), 6.

64. John Pierpont Morgan to James M. Brown, Jan. 12, 1882, document X123baf, Thomas A. Edison Papers Digital Edition, Rutgers University.

65. Frank L. Dyer, *Edison: His Life and Inventions* (New York: Harper and Brothers, 1910), 374.

66. See Matthew Josephson, *Edison: A Biography* (New York: McGraw-Hill, 1959), 255; Thomas Parke Hughes, *Networks of Power: Electrification in Western Society, 1880–1930* (Baltimore, MD: Johns Hopkins University Press, 1983), 70; Vaclav Smil, *Energy Transitions: History, Requirements, Prospects* (Santa Barbara, CA: Praeger, 2010), 41; Carroll W. Pursell, *Technology in America: A History of Individuals and Ideas*, 2nd ed. (Cambridge: MIT Press, 1990), 125. *Networks of Power* is one of the most comprehensive sources for information on the start-up and operation of the Pearl Street generating station, especially pages 70–73.

67. Hughes, *Networks of Power*, 43.

68. See T. A. Edison, System of electrical distribution, US Patent 274,290, issued Mar. 20, 1883. Edison had also organized and installed a three-wire experimental installation in Brockton, Massachusetts, in 1883 to test his three-wire system. William Lloyd Garrison, the well-known abolitionist, organized the Edison Electric Illuminating Company of Brockton in 1883 to introduce the new underground system to that city. For a breakdown of

the Brockton installation, see the testimony of W. J. Jenks in *Edison Electric Light Company v. F. P. Little Electrical Construction and Supply Company et al: On Letters Patent No. 281,576* (New York, 1896), 27.

69. "Edison's Newest Marvel."

70. The March 1883 article in *Engineering* is quoted in Hughes, *Networks of Power*, 91. "The Gaulard-Gibbs Secondary Generator," supplement, *Scientific American* 15, no. S387, June 2, 1883, 6172.

71. "Gaulard-Gibbs Secondary Generator," 6172.

72. For a more comprehensive description of alternating current versus direct current see R. Prasad, *Fundamentals of Electrical Engineering* (New Delhi: Prentice Hall, 2005), 1–60. See also Hughes, *Networks of Power*, 90–95.

73. Lucien Gaulard and John Dixon Gibbs, System of electric distribution, US Patent 351,589, issued Oct. 26, 1886.

74. Arnold Heertje and Mark Perlman, *Evolving Technology and Market Structure: Studies in Schumpeterian Economics* (Ann Arbor: University of Michigan Press, 1990), 129.

75. There are a number of sources that describe "the battle of the currents" in detail; see Hughes, *Networks of Power*, 106; and Jonnes, *Empires of Light*, 141–65.

76. For examples of how many members of the public viewed electricity as a panacea in miraculous and/or deistic terms see, Jacob L. Stone, *A Collection of Thoughts* (Chicago, 1881).

CHAPTER 6: ENERGY, UTOPIA, AND THE AMERICAN MIND

Epigraph: Henry Adams, *The Education of Henry Adams: An Autobiography* (Boston: Houghton Mifflin, 1918), 318.

1. Leo Marx, *The Machine in the Garden: Technology and the Pastoral Ideal in America* (New York: Oxford University Press, 1964), 232.

2. Roderick Frazier Nash, *Wilderness and the American Mind* (New Haven, CT: Yale University Press, 2001), 129.

3. G. Worthington, "Thirty Years of Electrical Supply in New York," *Electrical Review* 61, no. 11 (1912): 486.

4. One of the most comprehensive biographies of Nikola Tesla is W. Bernard Carlson, *Tesla: Inventor of the Electrical Age* (Princeton, NJ: Princeton University Press, 2013), 81.

5. Carlson, *Tesla*, 82.

6. Nikola Tesla, Electro magnetic motor, US Patent 382,279, issued May 1, 1888; Tesla, Electrical transmission of power, US Patent 382,280, issued May 1,

1888; Tesla, Electrical transmission of power, US Patent 382,281 issued May 1, 1888; Tesla, Method of converting and distributing electric currents, US Patent 382,282, issued May 1, 1888.

7. Stuart Banner, *The Death Penalty: An American History* (Cambridge, MA: Harvard University Press, 2003), 184. Banner's synopsis of the event on pages 184–89 of his book provides a concise summary of Kemmler's execution.

8. Thomas A. Edison, "The Dangers of Electric Lighting," *North American Review* 149 (Nov. 1889): 625.

9. See for example, "Met Death in the Wires Horrifying Spectacle on a Telegraph Pole: A Lineman Roasted in a Network of Wires—Thousands of People View the Terrible Sight," *New York Times*, Oct. 12, 1889; and "The Deadly Wires: An Old Man Receives an Electric Shock in Union Square," *New York Times*, Sept. 20, 1889.

10. "An Electrical Euclid," *Electric Power* 1, no. 9 (1889): 290.

11. "Far Worse than Hanging, Kemmler's Death Proves an Awful Spectacle," *New York Times*, Aug. 7, 1890.

12. Joseph P. Sullivan, "Fearing Electricity: Overhead Wire Panic in New York City," *IEEE Technology and Society Magazine* 14, no. 3 (1995): 7.

13. "Activity of the Westinghouse Electric and Manufacturing Company," *Electrical Engineer* 10 (1890): 404.

14. Benjamin Cummings Truman, *History of the World's Fair: Being a Complete and Authentic Description of the Columbian Exposition from Its Inception* (New York: E. B. Treat, 1893), 315.

15. Thomas Parke Hughes, *Networks of Power: Electrification in Western Society, 1880–1930* (Baltimore, MD: Johns Hopkins University Press, 1983), 105.

16. George Westinghouse Jr., "A Reply to Mr. Edison," *North American Review* 149, no. 397 (1889): 664.

17. "Mr. Edison Is Satisfied," *New York Times*, Feb. 21, 1892.

18. Robert W. Rydell, *All the World's a Fair: Visions of Empire at American International Expositions, 1876–1916* (Chicago: University of Chicago Press, 2013), 39.

19. "Pushing the Button," *Chicago Daily Tribune*, May 1, 1893.

20. Joseph M. Di Cola and David Stone, *Chicago's 1893 World's Fair* (Charleston, SC: Arcadia, 2012), 27.

21. Rydell, *All the World's a Fair*, 2.

22. Truman, *History of the World's Fair*, 315.

23. John Patrick Barrett, *Electricity at the Columbian Exposition: Including an Account of the Exhibits in the Electricity Building, the Power Plant in Machinery Hall, the Arc and Incandescent Lighting of the Grounds and Buildings* (Chicago: R. R. Donnelley, 1894), 99.

24. Carolyn Marvin, *When Old Technologies Were New: Thinking about Electric Communication in the Late Nineteenth Century* (New York: Oxford University Press, 1988), 172.

25. Trumbull White and William Igleheart, *The World's Columbian Exposition, Chicago, 1893* (Boston: J. K. Hastings, 1893), 316. A photograph of the exhibit that was recently on display at the Field Museum in Chicago may also be found on the same page.

26. Truman, *History of the World's Fair*, 609.

27. Teresa Dean, *White City Chips* (Chicago: Warren, 1895), 293.

28. Quoted in Keith Newlin, *Hamlin Garland: A Life* (Lincoln: University of Nebraska Press, 2008), 175.

29. Quoted in Reid Badger, *The Great American Fair: The World's Columbian Exposition and American Culture* (Chicago: Nelson Hall, 1979), 162.

30. "Electricity at the Fair," *Chicago Daily Tribune*, June 2, 1893.

31. Daniel Oscar Loy, *Poems of the White City* (Chicago, 1893), 84.

32. "Fair Electrified: Night Illumination of the Great White City," *Rock Island Daily Argus*, May 9, 1893.

33. "The Fairgrounds at Night," *Bismarck Weekly Tribune*, Apr. 28, 1893.

34. "Wizard Wonders at the Fair," *Omaha Daily Bee*, July 20, 1893, 2.

35. Dean, *White City Chips*, 370.

36. "Wizard Wonders at the Fair," 2.

37. Rossiter Johnson, *A History of the World's Columbian Exposition Held in Chicago in 1893* (New York: D. Appleton, 1897), 193.

38. Truman, *History of the World's Fair*, 318.

39. White and Igleheart, *World's Columbian Exposition*, 148. For information about the Standard Oil Company's Whiting Refinery see Alfred D. Chandler Jr. and Takashi Hikino, *Scale and Scope: The Dynamics of Industrial Capitalism* (Cambridge, MA: Belknap Press of Harvard University Press, 1994), 95.

40. White and Igleheart, *World's Columbian Exposition*, 148.

41. Adams, *Education of Henry Adams*, 384.

42. T. J. Jackson Lears, *No Place of Grace: Antimodernism and the Transformation of American Culture, 1880–1920* (Chicago: University of Chicago Press, 1994), 262. Lears addresses the debate that Adams was one who pro-

jected his personal and political disappointments "onto the cosmos," arguing instead that Adams expressed his ideas as an exercise in free intellectual play. Lears's chapter 7, "From Filial Loyalty to Religious Protest," expands upon this debate.

43. Henry G. Prout, *A Life of George Westinghouse* (New York: C. Scribner, 1922), 151.

44. "Storing Electricity," *Chicago Daily Tribune*, June 21, 1881.

45. "Storing Electricity."

46. "The New Source of Power," *New York Times*, Oct. 19, 1884.

47. Edward Dean Adams, *Niagara Power: History of the Niagara Falls Power Company, 1886–1918* (Niagara Falls, NY: Niagara Falls Power Company, 1927), 115–92.

48. Quoted in Patrick Vincent McGreevy, *Imagining Niagara: The Meaning and Making of Niagara Falls* (Amherst: University of Massachusetts Press, 1994), 107.

49. Nikola Tesla, "Tesla's Speech: The Age of Electricity," *Cassier's Magazine*, Mar. 1897, 381.

50. "Niagara's Mighty Force," *New York Tribune*, Aug. 22, 1897.

51. "Extensions to the Plant of the Niagara Falls Power Company," *American Gas Light Journal*, Jan. 17, 1898, 87.

52. "New Outlook," *New Outlook Magazine,* July 27, 1895, 128.

53. W. E. Curtis, "Almost Ready to Harness Niagara," *Scranton Tribune*, Aug. 29, 1896. See also "Distribution of Niagara's Power," *New York Times*, May 6, 1894; and "Niagara Put in Harness; Industrial Progress Claims the Great Cataract's Power," *New York Times*, July 7, 1895.

54. "Buffalo's Big Show," *Kansas City Journal*, Dec. 12, 1897. A copy of the Westinghouse advertisement can be seen in the *Engineer* (1898), iii.

55. Examples of terms or expressions related to electricity are not uncommon in the era; some examples include: "The world is electrified by the intelligence that Mr. Beecher wears a seal skin overcoat," in "Fashion and Gossip," *Salt Lake Herald*, Jan. 20, 1884; also, "Incidental to the play many realistic scenes including the Flying Railroad Train, Engine Tender and Cars, dashing through the tunnel, creating a furor, and received with cheers by the electrified audience," in "Pikes Matinee Today," *Cincinnnati Daily Star*, Mar. 18, 1880. A reference to James J. Hill, the Northern Pacific railroad magnate, describes Hill in a subheadline, "Has just gained control of the Northern Pacific. His habit of discharging employees. How he once caught a Tartar. A human dynamo," in "A Second Jay Gould," *Norfolk Virginian*, May 28, 1895.

56. "Literary Notices," *New-York Mirror*, July 21, 1832, 20.

57. Herman Melville, *Moby-Dick: Or, the Whale* (New York: Harper and Brothers, 1851), 560–61.

58. Mark Twain, *A Connecticut Yankee in King Arthur's Court* (New York: Harper and Brothers, 1889), 406–7.

59. Twain, *Connecticut Yankee*, 405, 28. For the Tesla reference see Margaret Cheney, *Tesla: Man out of Time* (New York: Simon and Schuster, 2001), 21.

60. Jean Pfaelzer, *The Utopian Novel in America, 1886–1896: The Politics of Form* (Pittsburgh, PA: University of Pittsburgh Press, 1985), 3.

61. Kenneth M. Roemer, *The Obsolete Necessity: America in Utopian Writings, 1888–1900* (Kent, OH: Kent State University Press, 1976), 156, 11.

62. Bellamy's work was the second-best-selling American novel in the nineteenth century behind Harriet Beecher Stowe's *Uncle Tom's Cabin*. Hundreds of "Bellamy societies" were formed in response to the work; see Ben Jackson, Marc Stears, and Michael Freeden, eds., *Liberalism as Ideology: Essays in Honour of Michael Freeden* (Oxford: Oxford University Press, 2012), 16.

63. Edward Bellamy, *Looking Backward: 2000–1887* (Boston: Ticknor, 1888), 165.

64. Edward Bellamy, *Equality* (New York: Appleton, 1898), 298–99, 347.

65. Bellamy, *Equality*, 236, 25.

66. Ignatius Donnelly, *Caesar's Column: A Story of the Twentieth Century* (Chicago: F. J. Schulte, 1890), 359.

67. Donnelly, *Caesar's Column*, 125.

68. Donnelly, *Caesar's Column*, 10, 22, 12.

69. Susan Harris Smith, *American Drama: The Bastard Art* (Cambridge: Cambridge University Press, 2006), 173.

70. William Dean Howells, *A Traveler from Altruria: Romance* (New York: Harper and Brothers, 1894), 284.

71. King Camp Gillette, *The Human Drift* (Boston: New Era, 1894), 88–90.

72. Mary E. Bradley Lane, *Mizora: A World of Women* (Lincoln: University of Nebraska Press, 1999), 75.

73. Anna Bowman Dodd, *The Republic of the Future, or, Socialism a Reality* (New York: Cassell, 1887), 15.

74. Dodd, *Republic of the Future*, 27.

75. Howells, *Traveler from Altruria*, 260.

CHAPTER 7: TURBINES, COAL, AND CONVENIENCE

Epigraph: Walter J. Ballard, "Industrial and Commercial Progress since 1870," *Protectionist*, July 1902, 160.

1. Julie Wernau, "Closure of Chicago's Crawford, Fisk Electric Plants Ends Coal Era," *Chicago Tribune*, Aug. 30, 2012.

2. "One of the World's Seven Wonders," *Chicago Daily Tribune*, Dec. 6, 1903.

3. For electrification statistics see United States Bureau of the Census, *Historical Statistics of the United States, Colonial Times to 1957; a Statistical Abstract Supplement* (Washington, DC: US Government Printing Office, 1960), 510. Household electrification statistics saw a significant rise early in the century—in 1917, 24 percent of US dwellings were electrified, and the number had doubled by 1920 to 47 percent. See also Ruth Schwartz Cowan, "The 'Industrial Revolution' in the Home: Household Technology and Social Change in the 20th Century," *Technology and Culture* 17, no. 1 (1976): 5.

4. Thomas Parke Hughes, *Networks of Power: Electrification in Western Society, 1880–1930* (Baltimore, MD: Johns Hopkins University Press, 1983), 79.

5. Hughes, *Networks of Power*, 79.

6. Walt Patterson, *Keeping the Lights On: Towards Sustainable Electricity* (Baltimore, MD: Brookings Institution Press, 2012), 45–46. Patterson provides an explanation of Edison's rate structure. For an explanation of the rates that Edison charged in New York, see Thomas C. Martin, *Forty Years of Edison Service, 1882–1922: Outlining the Growth and Development of the Edison System in New York City* (New York: New York Edison Company, 1922), 139.

7. Vaclav Smil, *Energy Transitions: History, Requirements, Prospects* (Santa Barbara, CA: Praeger, 2010), 52–53.

8. Smil, *Energy Transitions*, 54. See also John F. Wasik, *The Merchant of Power: Samuel Insull, Thomas Edison, and the Creation of the Modern Metropolis*, 1st ed. (New York: Palgrave Macmillan, 2006), 84.

9. Hughes, *Networks of Power*, 211.

10. Philip Torchio, "Discussion on the Present Limits of Speed and Power of Single Shaft Steam Turbines," *Transactions of the American Institute of Electrical Engineers* 38, no. 2 (1919): 1547.

11. Hughes, *Networks of Power*, 212.

12. Wasik, *Merchant of Power*, 57.

13. Wasik, *Merchant of Power*, 57.

14. Forrest McDonald, *Insull* (Chicago: University of Chicago Press, 1962), 62.

15. W. M. Walbank, "Lachine Rapids Plant and the Cost of Producing Power for Generating Electricity Therefrom," in *Proceedings of the Twenty-First National Electric Light Association Convention*, ed. National Electric Light Association Convention (New York, 1898), 257–76.

16. Walbank, "Lachine Rapids Plant," 263.

17. Wasik, *Merchant of Power*, 84.

18. Hughes, *Networks of Power*, 211.

19. McDonald, *Insull*, 100; and "Fisk Street Station of the Commonwealth Edison Company, Chicago," *Western Electrician* 38, no. 1 (Jan. 6, 1906): 1.

20. McDonald, *Insull*, 100; and "Fisk Street Station," 1.

21. United States Bureau of the Census, *Central Electric Light and Power Stations* (Washington, DC: US Government Printing Office, 1910), 94.

22. James McNerney, Jessika E. Trancik, and J. Doyne Farmer, "Historical Costs of Coal-Fired Electricity and Implications for the Future," Santa Fe Institute Working Paper 2009-12-047 (Santa Fe, NM: Dec. 16, 2009), 3–7.

23. Paul M. Lincoln, "Modern Developments in Power Generation," *Electrical Review and Western Electrician* 64, no. 24 (June 27, 1914): 1302.

24. Power plants fueled by petroleum gas, lignite, natural gas, and diesel were in operation in the first decades of the 1900, but they were the exception. Examples include a small-scale gasoline plant in Stephenville, Texas, in 1911, a three-megawatt lignite plant in Blooming Grove, Texas, in 1912, a seven-megawatt natural gas plant in Independence, Kansas, in 1912, and a small diesel plant in Lebanon, Indiana, in 1904. See "Technical Aspects of the Period," in *Central Electric Light and Power Stations and Street and Electrical Railways: With Summary of the Electrical Industries 1912*, ed. W. M. Steuart, T. C. Martin, F. L. Sanford, and S. B. Ladd (Washington, DC: United States Bureau of the Census, 1915), 120–21.

25. United States Department of Agriculture, "Comparative Cost of Power from Different Sources," *Farmers' Bulletin*, no. 277, 1908, 7.

26. Lorin E. Imlay, Lyle A. Whitsit, B. J. Peterson, et al., "Appendix G: Hydroelectric Plants for the Superpower System," in *Copper Deposits of the Tyrone District, New Mexico*, by Sidney Paige, 192–203, United States Geological Survey, Professional Paper No. 122 (Washington, DC: United States Government Printing Office, 1922), 192.

27. "Hydro Electric Development," *Power* 47, no. 2 (June 30, 1918): 237.

28. D. A. Shearer, "Coal-Pit Mine Mouth Power Plants," *Power* 47, no. 2 (June 30, 1918): 256.

29. McDonald, *Insull*, 108.

30. John Moody, *Moody's Manual of Investments: American and Foreign* (New York: Moody's Investors Service, 1922), 349.

31. McDonald, *Insull*, 110. In preparation for threatened strikes, Insull accumulated large reserves of coal at several power plants including the Fisk Street Station. In the months prior to April in 1914, for example, Commonwealth Edison stockpiled four hundred thousand tons of coal as a reserve; see Samuel Insull, *Central-Station Electric Service; Its Commercial Development and Economic Significance as Set Forth in the Public Addresses (1897–1914) of Samuel Insull* (Chicago: Privately printed, 1915), 331; see also "Effect of Strike upon Central-Station Supply to Coal Mines," *Electrical Review* 75 (Nov. 29, 1919): 903.

32. Price V. Fishback, *Soft Coal, Hard Choices: The Economic Welfare of Bituminous Coal Miners, 1890–1930* (New York: Oxford University Press, 1992), 19–21. For relative oil and petroleum prices see Bruce Podobnik, *Global Energy Shifts* (Philadelphia: Temple University Press, 2008).

33. United States Bureau of the Census, *Central Electric Light and Power Stations and Street and Electrical Railways*, 334.

34. Hughes, *Networks of Power*, 363–64.

35. Hughes, *Networks of Power*, 119–21.

36. Articles of Incorporation, 1911, Samuel Insull Papers, series 8, boxes 49–55, Loyola University, Chicago.

37. Samuel Insull, "The Progress of Economic Power Generation and Distribution," *Journal of the American Society of Mechanical Engineers* 38 (1916): 845.

38. R. J. Bufard of General Electric to Frank C. Bosler, Apr. 24, 1913, box 47, folder 403, Bosler Collection, American Heritage Center, University of Wyoming, Laramie.

39. E. R. Reginald of the Electric Storage Battery Company to Frank C. Bosler, May 1, 1913, box 47, folder 403, Bosler Collection, American Heritage Center, University of Wyoming, Laramie.

40. United States Bureau of the Census, *Central Electric Light and Power Stations and Street and Electrical Railways*, 34.

41. Hughes, *Networks of Power*, 208.

42. United States Bureau of the Census, *Historical Statistics of the United*

States, Colonial Times to 1970 (Washington, DC: US Government Printing Office, 1975), 822.

43. United States Bureau of the Census, *Central Electric Light and Power Stations and Street and Electrical Railways*, 34.

44. United States Bureau of the Census, *Central Electric Light and Power Stations and Street and Electrical Railways*, 29.

45. United States Bureau of the Census, *Historical Statistics of the United States, Colonial Times to 1970*, 820. This information was derived from the statistical figures in the table titled "Net Production of Electric Energy, by Electric Utility and Industrial Generating Plants, by Type of Plant: 1902 to 1970, Series S 32–43." For this analysis, the total reported output in kilowatt-hours was compared to total share of generating plant type, hydropower versus steam.

46. "Generation of Electricity in the Largest Steam-Driven Station," *Electrical Review and Western Electrician* 57, no. 23 (1910): 1124. For the *Chicago Tribune* quote see "One of the World's Seven Wonders."

47. David M. Myers, *The Power Plant* (New York: Industrial Extension Institute, 1922), 466.

48. Clarence C. Brinley, "Reducing Cost with Mechanical Stokers," *Engineering Magazine*, Nov. 1915, 277.

49. "A Swamp Becomes a Busy Scene," *St. Paul Globe*, Aug. 30, 1903, 27.

50. David Stradling, *Conservation in the Progressive Era: Classic Texts* (Seattle: University of Washington Press, 2004), 73.

51. "Smoke Not Necessary, Experts Say Plants Can Be Operated without Producing It," *New York Tribune*, Apr. 22, 1909.

52. "Uncle Sam's Coal Is Being Wasted, Report Declares," *Washington Times*, Jan. 1, 1913.

53. D. T. Randall and H. W. Weeks, *The Smokeless Combustion of Coal in Boiler Plants: With a Chapter on Central Heating Plants*, vol. 2 (Washington, DC: US Government Printing Office, 1909), 184.

54. Michael D. Vanden Berg, *Annual Review and Forecast of Utah Coal Production and Distribution—2008* (Salt Lake City: Utah Geological Survey, 2010), 15.

55. Robert H. Orrkok and George A. Fernald, *Engineering of Power Plants* (New York: McGraw-Hill, 1921), 195.

56. United States Department of Energy, Energy Information Administration, *International Energy Outlook 2013 with Projections to 2040* (Washington, DC: US Government Printing Office, 2013), 112.

57. William R. Leach, *Land of Desire: Merchants, Power, and the Rise of a New American Culture* (New York: Knopf Doubleday, 1993), xiii.

58. Leach, *Land of Desire*, 367.

59. Leach, *Land of Desire*, 368.

60. "Watt's What," *National Electric Light Association Bulletin* 7, no. 7 (1920): 523.

61. David F. Noble, *America by Design: Science, Technology, and the Rise of Corporate Capitalism* (Oxford: Oxford University Press, 1979), 18.

62. Harold L. Platt, *The Electric City: Energy and the Growth of the Chicago Area, 1880–1930* (Chicago: University of Chicago Press, 1991), 155.

63. John Craig Hammond, "New Business for Electric Central Stations," *Cassier's Magazine*, vol. 30, no. 3, July 1906, 259.

64. Platt, *Electric City*, 237.

65. Mildred Maddocks, "We Recommend Electricity," *Good Housekeeping*, June 1918, 57.

66. Ruth Schwartz Cowan, *More Work for Mother: The Ironies of Household Technology from the Open Hearth to the Microwave* (New York: Basic Books, 1983), 91; and Janice Williams Rutherford, *Selling Mrs. Consumer: Christine Frederick and the Rise of Household Efficiency* (Athens: University of Georgia Press, 2010), 1. Rutherford's work focuses on "the cult of efficiency" and efficiency expert Christine Frederick in the second and third decades of the twentieth century. Frederick encouraged women to become "professional consumers" by becoming more efficient in their "traditional domestic sphere."

67. "Report of National Commodity Advertisers' Division Advertising and Publicity Service Bureau," *National Electric Light Association Bulletin* 7, no. 7 1920, 554. The August 1920 *NELA Bulletin* listed a variety of magazines including *Collier's, Country Gentleman, Successful Farming, Popular Mechanics, Farm & Fireside, Popular Science, Youth's Companion, American Boy, Hoard's Dairyman*, and *National Geographic*, to name a few.

68. Francis Arnold Collins, "Pops of Popular Science," *Boys' Life*, Aug. 1920, 38; and "Turn the Switch and Heat Your Bath," *Popular Mechanics*, Feb. 1920, 50.

69. "Do You Glow with Health?" *Popular Mechanics*, Mar. 1922, 177.

70. Samuel H. Monell, *Electricity in Health and Disease; a Treatise of Authentic Facts for General Readers, in Which Is Shown How Electric Currents Are Made to Act as Curative Remedies, Together with an Account of the Principal Diseases Which Are Benefited by Them* (New York: McGraw, 1907), 166, 225.

71. Monell, *Electricity in Health and Disease*, v.

72. George W. Jacoby and J. Ralph Jacoby, *Electricity in Medicine; a Practical Exposition of the Methods and Use of Electricity in the Treatment of Disease, Comprising Electrophysics, Apparatus, Electrophysiology and Electropathology, Electrodiagnosis and Electroprognosis, Generalelectrotherapeutics and Special Electrotherapeutics* (Philadelphia: P. Blakiston's Son, 1919), 470.

73. Carolyn Thomas de la Peña, *The Body Electric* (New York: NYU Press, 2005).

74. Ronald C. Tobey, *Technology as Freedom: The New Deal and the Electrical Modernization of the American Home* (Berkeley: University of California Press, 1996), 28.

75. United States Bureau of the Census, *Historical Statistics of the United States, Colonial Times to 1970*, 827.

76. Helen L. Bartlett, "The Efficiency of Electricity to the Housekeeper," *Electrical Review and Western Electrician* 64, no. 19 (1914): 934.

77. Chicago Department of Health, *Bulletin of Chicago School of Sanitary Instruction* (Chicago: Chicago Department of Health, 1920), 99.

78. Joseph K. Hart, "Neighborhood and Community Life," paper presented at the National Conference of Social Work, Cleveland, OH, 1926, accessed Oct. 8, 2014, http://quod.lib.umich.edu/n/ncosw/ACH8650.1926.001?rgn=main;view=fulltext.

79. "Clean Electric Motors," *Popular Mechanics*, Sept. 1, 1924, 122; "Clean Electric Plants," *Motor Boating*, Dec. 1, 1923, 98; "Clean Electric Light," *Country Gentleman*, July 6, 1918, 25.

80. United States Bureau of the Census, *Historical Statistics of the United States, Colonial Times to 1970*, 826.

CONCLUSION

1. Frank Bohn, "The Electric Age: A New Utopia," *New York Times*, Oct. 2, 1927, 1.

2. Mrs. W. H. Stewart, "Letter Writing Contest," *Washington Times*, May 15, 1922, 14.

3. Ruth Schwartz Cowan, *More Work for Mother: The Ironies of Household Technology from the Open Hearth to the Microwave* (New York: Basic Books, 1983), 169–73. See also Daniel Nelson, *Farm and Factory: Workers in the Midwest, 1880–1990* (Bloomington: Indiana University Press, 1995), 85; and United States Bureau of the Census, *Historical Statistics of the United States, Colonial Times to 1970* (Washington, DC: US Government Printing Office, 1975), 827.

4. Cowan, *More Work for Mother*, 169–73; and Nelson, *Farm and Factory*, 85.

5. David E. Nye, *Electrifying America: Social Meanings of a New Technology, 1880–1940* (Cambridge: MIT Press, 1990), 268.

6. Charles Robert Gibson, *The Romance of Modern Electricity: Describing in Non-Technical Language, What Is Known about Electricity and Many of Its Interesting Applications* (Philadelphia: J. B. Lippincott, 1906), 339.

7. Lewis Mumford, *Technics and Civilization* (1934; repr., Chicago: University of Chicago Press, 2010), 223.

8. Mumford, *Technics and Civilization*, 155–57.

9. For an example, see George Iles, "Why Progress Is by Leaps," *Popular Science*, June 1896, 226. Iles argues, "Were electricity unmastered there could be no democratic government in the United States," and due to electrification, "today the drama of national affairs is more directly in the view of every American citizen than a century ago." See also General Electric advertising from 1925, "Abundant Electricity," in *Michigan Alumnus*, Mar. 1920, 707.

10. Herbert Hoover, *Public Papers of the Presidents of the United States, Mar. 4 to Dec. 31, 1929* (Washington, DC: Office of the Federal Register, National Archives and Records Service, 1929), 502, accessed Nov. 25, 2015, http://name.umdl.umich.edu/4731615.1929.001.

11. United States Bureau of the Census, *Historical Statistics of the United States, Colonial Times to 1970*, 826.

12. Franklin D. Roosevelt, "Campaign Address in Portland, Oregon on Public Utilities and Development of Hydro-Electric Power," Sept. 21, 1932, online by Gerhard Peters and John T. Woolley, *The American Presidency Project*, accessed Oct. 8, 2015, http://www.presidency.ucsb.edu/ws/?pid=88390.

13. Thomas Parke Hughes, *Networks of Power: Electrification in Western Society, 1880–1930* (Baltimore, MD: Johns Hopkins University Press, 1983), 364.

14. Tim Culvahouse, *The Tennessee Valley Authority: Design and Persuasion* (New York: Princeton Architectural Press, 2007), 103.

15. United States Bureau of the Census, *Historical Statistics of the United States, Colonial Times to 1970*, 826.

16. Dwight D. Eisenhower, "Address Before the General Assembly of the United Nations on Peaceful Uses of Atomic Energy, New York City," in *Public Papers of the Presidents of the United States. Containing the Public*

Messages, Speeches, and Statements of the President: January 20 to December 31, 1953 (Washington, DC: US Government Printing Office, 1955).

17. Alan C. Fisher Jr., "You and the Obedient Atom," *National Geographic*, vol. 114, no. 3, Sept. 1958, 303.

18. George Basalla, *The Evolution of Technology* (Cambridge: Cambridge University Press, 1988), 6.

19. Basalla, *Evolution of Technology*, 13.

20. John A. Etzler, *The Paradise within the Reach of All Men, without Labor, by Powers of Nature and Machinery: An Address to All Intelligent Men* (Pittsburgh, PA: Etzler and Reinhold, 1833), 3–4.

21. Sir William Thomson, *Reflections on the Motive Power of Heat and on Machines Fitted to Develop That Power, from the Original French of N. L. S. Carnot* (New York: John Wiley and Sons, 1890), 225, 44.

22. Steven Stoll, *The Great Delusion: A Mad Inventor, Death in the Tropics, and the Utopian Origins of Economic Growth* (New York: Hill and Wang, 2008), 53; John A. Etzler, *Two Visions of J. A. Etzler . . . A Revelation of Futurity* (Surrey, 1844), 7.

23. Abraham Lincoln, *Collected Works of Abraham Lincoln*, ed. Roy P. Basler, vol. 2, *1848–1858* (New Brunswick, NJ: Rutgers University Press, 1953), 442.

24. William Conant Church, *The Life of John Ericsson*, vol. 2 (London: Sampson, Low, Marston, 1892), 265.

25. W. H. Krause, *Ericsson Cycle Gas Turbine Powerplants* (Santa Monica, CA: RAND Corporation, 1979), v, accessed Oct. 8, 2015, http://www.rand.org/pubs/reports/R2327.

26. W. Bernard Carlson, *Tesla: Inventor of the Electrical Age* (Princeton, NJ: Princeton University Press, 2013), 84.

27. Nikola Tesla, "Our Future Motive Power," *Everyday Science and Mechanics*, Dec. 1931, 230–36.

28. Anna Hirtenstein, "Makai Builds Ocean Thermal-Energy Demo Plant in Hawaii," *Bloomberg Business News*, Aug. 21, 2015, accessed Oct. 8, 2015, http://www.bloomberg.com/news/articles/2015-08-21/makai-builds-ocean-thermal-energy-demo-plant-in-hawaii.

29. "Mr. Brush's Windmill Dynamo," *Scientific American* 63, no. 25 (Dec. 20, 1890): 389.

30. Robert W. Righter, *Wind Energy in America: A History* (Norman: University of Oklahoma Press, 1996), 53.

31. R. J. Bufard of General Electric to Frank C. Bosler, Apr. 24, 1913, box 47, folder 403, Bosler Collection, American Heritage Center, University of Wyoming, Laramie.

32. James Rodger Fleming, *Historical Perspectives on Climate Change* (New York: Oxford University Press, 2005), 65.

33. Spencer R. Weart, *The Discovery of Global Warming* (Cambridge, MA: Harvard University Press, 2003), 1.

34. Paul Sabatier, "An Advocacy Coalition Framework of Policy Change and the Role of Policy-Oriented Learning Therein," *Policy Sciences* 21, no. 2–3 (1988): 131.

35. Sabatier, "Advocacy Coalition Framework," 129.

36. Joseph Blumberg, "The Enduring Legacy of Ecologist Herbert Bormann," *Dartmouth Now*, June 28, 2012, accessed Nov. 9, 2015, http://now.dartmouth.edu/2012/06/enduring-legacy-ecologist-herbert-bormann.

37. Chris C. Park, *Acid Rain: Rhetoric and Reality* (New York: Routledge, 2013), 28.

38. Stephen Ansolabehere and David M. Konisky, *Cheap and Clean: How Americans Think about Energy in the Age of Global Warming* (Cambridge: MIT Press, 2014), 44.

39. Committee on Electricity in Economic Growth, *Electricity in Economic Growth* (Washington, DC: National Academy Press, 1986), 41.

40. Jimmy Carter, "Remarks in Golden, Colorado, May 3, 1978," in *Public Papers of the Presidents of the United States, Book 1—January 1 to June 30, 1978* (Washington, DC: US Government Printing Office, 1979), 825.

41. David Biello, "Where Did the Carter White House's Solar Panels Go?" *Scientific American*, Aug. 6, 2010, accessed Oct. 8, 2015, http://www.scientificamerican.com/article/carter-white-house-solar-panel-array/.

42. Benjamin K. Sovacool, *Contesting the Future of Nuclear Power: A Critical Global Assessment of Atomic Energy* (Hackensack, NJ: World Scientific, 2011), 2; and Carter N. Lane, *Acid Rain: Overview and Abstracts* (New York: Nova Science, 2003), 10.

43. Kevin Coyle, *Environmental Literacy in America: What Ten Years of NEETF/Roper Research and Related Studies Say About Environmental Literacy in the U.S.* (Washington, DC: National Environmental Education and Training Foundation, 2005), v.

44. B. G. Southwell, J. J. Murphy, J. E. DeWaters, and P. A. LeBaron, *Americans' Perceived and Actual Understanding of Energy*, RTI Press publi-

cation No. RR-0018-1208 (Research Triangle Park, NC: RTI Press, 2012), 5, accessed Nov. 15, 2015, http://www.rti.org/rtipress.

45. From the author's surveys of approximately 350 undergraduate students from 2000 through 2015 at the University of Toledo.

46. Pavel Somavat and Vinod Namboodiri, "Energy Consumption of Personal Computing Including Portable Communication Devices," *Journal of Green Engineering*, July 2012, 447–475, 467.

47. International Energy Agency, *CO_2 Emissions from Fuel Combustion, 2015 Preliminary Edition*, accessed Nov. 1, 2015, http://www.iea.org/publications/freepublications/publication/co2-emissions-from-fuel-combustion-for-oecd-countries---2015-preliminary-edition---factsheet.html.

48. Walt Kelly, *Pogo: We Have Met the Enemy and He Is Us* (New York: Simon and Schuster, 1972), 7.

49. Adam Frank, "Climate Change Is Not Our Fault," *Cosmos & Culture*, *NPR*, Oct. 28, 2015, accessed Nov. 10, 2015, http://www.npr.org/sections/13.7/2015/10/06/446109168/climate-change-is-not-our-fault.

REFERENCES

MANUSCRIPT COLLECTIONS

Articles of Incorporation. Samuel Insull Papers, Series 8, Boxes 49–55. Loyola University Chicago Archives, Chicago.

Bosler Collection. American Heritage Center, University of Wyoming, Laramie.

Thomas A. Edison Papers Digital Edition. Rutgers University, New Brunswick, NJ.

Xanthus Smith to Sophie Dupont, June 6, 1876. Group 9, Xanthus Smith Collection, Wintherthur Manuscripts. Hagley Museum and Library, Wilmington, DE.

MUNICIPAL GOVERNMENT DOCUMENTS

Chicago Department of Health. *Bulletin of Chicago School of Sanitary Instruction.* Edited by Chicago School of Sanitary Instruction. Chicago: Chicago Department of Health, 1920.

Chicago Department of Health. *Report of the Dept. of Health of the City of Chicago.* 1891. Cleveland Division of Health. *Annual Report.* 1881.

New York Department of Health and New York Board of Health. *Annual Report of the Board of Health of the Department of Health of the City of New York for the Year Ending.* 1872.

US GOVERNMENT DOCUMENTS

Carter, Jimmy. *Public Papers of the Presidents of the United States. Book 1—January 1 to June 30, 1978.* Washington, DC: US Government Printing Office, 1979.

Eisenhower, Dwight D. *Public Papers of the Presidents of the United States. Containing the Public Messages, Speeches, and Statements of the President: January 20 to December 31, 1953.* Washington, DC: US Government Printing Office, 1955.

Hoover, Herbert. *Public Papers of the Presidents of the United States, March 4 to December 31, 1929.* Washington, DC: Office of the Federal Register, National Archives and Records Service, 1929. Accessed November 25, 2015. http://name.umdl.umich.edu/4731615.1929.001.

Imlay, Lorin E., Lyle A. Whitsit, B. J. Peterson, et al. "Appendix G: Hydroelectric Plants for the Superpower System." In *Copper Deposits of the Tyrone District, New Mexico,* by Sidney Paige, 192–203. United States Geological Survey Professional Paper No. 122. Washington, DC: US Government Printing Office, 1922.

United States Bureau of the Census. *Central Electric Light and Power Stations.* Washington, DC: US Government Printing Office, 1910.

United States Bureau of the Census. *Central Electric Light and Power Stations and Street and Electrical Railways: With Summary of the Electrical Industries 1912.* Edited by W. M. Steuart, T. C. Martin, F. L. Sanford, and S. B. Ladd. Washington, DC: US Government Printing Office, 1915.

United States Bureau of the Census. *Historical Statistics of the United States 1789–1945.* Washington, DC: Government Printing Office, 1949.

United States Bureau of the Census. *Historical Statistics of the United States, Colonial Times to 1957; a Statistical Abstract Supplement.* Washington, DC: US Government Printing Office, 1960.

United States Bureau of the Census. *Historical Statistics of the United States, Colonial Times to 1970.* Washington, DC: US Government Printing Office, 1975.

United States Centennial Commission. *International Exhibition, 1876.* Edited by A. T. Goshorn, Francis Walker, and Dorsey Gardner. Washington, DC: US Government Printing Office, 1880.

United States Centennial Commission. *Reports and Awards.* Philadelphia: J. B. Lippincott, 1877.

United States Department of Agriculture. "Comparative Cost of Power from Different Sources." *Farmers' Bulletin,* no. 277, 1908, 7.

United States Department of Energy, Energy Information Administration. *International Energy Outlook 2013 with Projections to 2040.* Washington, DC: US Government Printing Office, 2013.

United States House of Representatives. *House Documents.* Washington, DC: US Government Printing Office, 1844.

INTERNATIONAL DOCUMENTS

International Energy Agency. CO_2 *Emissions from Fuel Combustion, 2015 Preliminary Edition.* Accessed November 1, 2015. http://www.iea.org/publications/freepublications/publication/c02-emissions-from-fuel-combustion-for-oecd-countries—2015-preliminary-edition—fact sheet.html.

US PATENTS

Aikman, Peter A. Stove damper. US Patent 233,456, issued October 19, 1880.

Allen, Stephen M. Chimney cowl. US Patent 2,568, issued April 21, 1842.

Andersen, Hans J. Heater and ventilator. US Patent 235,486, issued December 14, 1880.

Arnoux, Antoine. Gas burner (to minimize smoke). US Patent 764, issued June 4, 1838.

Austin, Julius. Wind wheel. US Patent 231,253, issued August 17, 1880.

Aylsworth, Chadiah. Improvement in water-wheels. US Patent 3,959, issued March 21, 1845.

Bain, David. Chimney cowl. US Patent X7,409, issued February 5, 1833.

Baldwin, James F. Ventilator for dwellings. US Patent 233,962, issued November 2, 1880.

Baldwin, Matthias W. Art of managing and supplying fire for generating steam in locomotive engines. US Patent 54, issued October 15, 1836.

Bales, Isaac. Combined house ventilator and register. US Patent 232,166, issued September 14, 1880.

Barnes, Charles T. Smoke and gas consumer for stoves. US Patent 231,142, issued August 17, 1880.

Bissell, William C. P. Smoke and gas consuming furnace. US Patent 234,385, issued November 16, 1880.

Blewett, George L. Spark-extinguisher. US Patent 225,326, issued March 9, 1880.

Brink, John Robert. Wind engine. US Patent 232,673, issued September 28, 1880.

Brown, Huntington. Smoke-stack. US Patent 226,439, issued April 13, 1880.

Bryan, Oliver. Hot-air furnace. US Patent 234,232, issued November 9, 1880.

Campfield, Hampton E. Spark-arrester. US Patent 228,442, issued June 8, 1880.

Caywood, Charles C. Chimney cap and cowl. US Patent 224,873, issued February 24, 1880.

Cooper, George B. F. Spark-arrester. US Patent 225,572, issued March 16, 1880.

Craig, Rufus S. Spark-arrester. US Patent 233,400, issued October 19, 1880.

Croft, Henry, Sr., and Henry Croft Jr. Wind wheel. US Patent 224,817, issued February 24, 1880.

Dunn, Edward. Spark-arrester. US Patent 227,420, issued May 11, 1880.

Edison, T. A. Electric lamp. US Patent 223,898, issued January 27, 1880.

Edison, T. A. System of electrical distribution. US Patent 274,290, issued March 20, 1883.

Etzler, John A. Navigating and propelling vessels by the action of the wind and waves. US Patent 2,533, issued April 1, 1842.

Felton, Benjamin W. Ventilator for chimney-caps. US Patent 226,507, issued April 13, 1880.

Garatt, John F. Windmill. US Patent 234,975, issued November 30, 1880.

Gaulard, Lucien, and John Dixon Gibbs. System of electric distribution. US Patent 351,589, issued October 26, 1886.

Gilstrap, Jacob. Wind wheel. US Patent 236,018, issued December 28, 1880.

Gray, T. B. Wind wheel. US Patent 232,815, issued October 5, 1880.

Gray, William. Furnace. US Patent 233,614, issued October 26, 1880.

Grosebeck, David. Spark-arrester. US Patent 235,762, issued December 21, 1880.

Gunther, George A. Locomotive spark-extinguisher. US Patent 234,274, issued November 9, 1880.

Harris, J. Nelson. Grate and fender for fire-places. US Patent 228,994, issued June 2, 1880.

Henderson, Leonard. Smoke and dust arrester for railway cars. US Patent 225,448, issued March 9, 1880.

Horton, N. N. Heater, cooler, and ventilator for railroad cars. US Patent 227,977, issued May 25, 1880.

Hubbard, Noahdiah W. Improvement in current water-wheels, being a plan for giving increased power to such wheels. US Patent 2,027, issued April 2, 1841.

Hurd, Joseph. Cap for regulating draft of chimneys. US Patent 3,854, issued December 12, 1844.

Jackson, Andrew T. Chimney flue and shield. US Patent 226,074, issued March 30, 1880.

Johnson, William E. Fire-safe for smoke-houses. US Patent 226,861, issued April 27, 1880.

Kendel, Adolphus C. Smoke-bell. US Patent 231,587, issued August 24, 1880.

LaFrance, Asa W. Smoke-stack. US Patent 227,272, issued May 4, 1880.

Lamoureux, Chauncey. Chimney cap. US Patent 230,483, issued July 27, 1880.

Lange, Urban B. A. Apparatus for preventing chimneys from smoking. US Patent X8,453, issued October 14, 1834.

Lewis, William, and Thomas J. Lewis. Improvement in horizontal windmills. US Patent 583, issued January 27, 1838.

Lloyd, John S. Smoke and cinder conveyor for locomotives. US Patent 227,550, issued May 11, 1880.

Makely, Jacob. Improvement in windmills. US Patent 479, issued November 23, 1837.

Mark, John W. Ventilator. US Patent 230,952, issued August 10, 1880.

McNiel, Elias H. Steam-boiler and furnace. US Patent 225,625, issued March 16, 1880.

Mitchell, Alexander. Cinder-guard. US Patent 226,869, issued April 27, 1880.

Morse, Samuel F. B. Improvement in the mode of communicating information by signals by the application of electro-magnetism. US Patent 1,647, issued June 20, 1840.

Mott, Jordon. Improvement for chimney caps. US Patent 2,887, issued December 17, 1842.

Myers, George A. Wind wheel. US Patent 229,907, issued July 13, 1880.

Neuhaus, Karl W. Spark-arrester. US Patent 228,922, issued June 15, 1880.

Perky, Francis A. Spark-arrester. US Patent 153,907, reissued August 31, 1880.

Pohl, Anton. Spark-arrester. US Patent 230,568, issued July 27, 1880.

Preston, J. C. Wind wheel. US Patent 232,205, issued September 14, 1880.

Pursell, Harry. Grate-front. US Patent 225,074, issued March 2, 1880.

Putnam, John P. Ventilating-gasolier. US Patent 233,372, issued October 19, 1880.

Quinn, Patrick. Upright steam-boiler. US Patent 226,880, issued April 27, 1880.

Rice, Augustus. Chimney cap. US Patent 7,275, issued April 9, 1850.

Rideout, Alexander. Warming and ventilating. US Patent 229,842, issued July 13, 1880.

Ridley, Henry A. Spark-arrester. US Patent 233,022, issued October 5, 1880.

Robbins, Edward, Jr., and William Ashby. Construction of water-wheels. US Patent 1,525, issued March 25, 1840.

Roberts, George S. Smoke-bell support for gas fixtures. US Patent 230,060, issued July 13, 1880.

Rolph, Benjamin M. Windmill. US Patent 234,204, issued November 9, 1880.

Russell, James M. Spark-arrester. US Patent 229,642, issued July 6, 1880.

Sampsel, John E. Smoke-box and stack for locomotive. US Patent 227,657, issued May 18, 1880.

Simpson, James. Construction of smoke stacks of locomotive or stationary steam engines and other chimneys for preventing the escape of sparks. US Patent 161, issued April 17, 1837.

Sinton, David. Smoke-consuming furnace. US Patent 233,168, issued October 12, 1880.

Smith, H. B. Wind wheel. US Patent 232,558, issued September 21, 1880.

Southworth, F. H. Tide or current wheel. US Patent 1,478, issued January 23, 1840.

Staples, Arthur. Smoke-stack. US Patent 230,673, issued August 3, 1880.

Strait, Ransom E. Wind engine. US Patent 225,539, issued March 16, 1880.

Suitt, James. Spark-arrester. US Patent 224,802, issued February 24, 1880.

Sumner, Palmer. Chimney cowl. US Patent 2,964, issued February 2, 1843.

Tesla, Nikola. Electrical transmission of power. US Patent 382,280, issued May 1, 1888.

Tesla, Nikola. Electrical transmission of power. US Patent 382,281, issued May 1, 1888.

Tesla, Nikola. Electro magnetic motor. US Patent 382,279, issued May 1, 1888.

Tesla, Nikola. Method of converting and distributing electric currents. US Patent 382,282, issued May 1, 1888.

Thompson, Otis D. Wind wheel. US Patent 235,470, issued December 14, 1880.

Thorton, William M. Spark-arrester. US Patent 229,207, issued June 22, 1880.

Timlin, David J. Spark-arrester. US Patent 234,349, issued November 9, 1880.

Wiggin, John E. Spark-arrester. US Patent 224,497, issued February 10, 1880.

Wiggin, John E. Spark-arrester. US Patent 226,378, issued April 6, 1880.

Woodward, Abijah. Improvement in tub water-wheels. US Patent 1,589, issued May 8, 1840.

Wright, Pratt. Spark-arrester for locomotives. US Patent 228,431, issued June 1, 1880.
Zeck, Michael. Spark-arrester. US Patent 223,427, issued January 6, 1880.
Zimmerman, William. Improvement in wind wheels. US Patent 2,107, issued May 29, 1841.

PUBLISHED SOURCES

"Activity of the Westinghouse Electric and Manufacturing Company." *Electrical Engineer* 10 (1890): 404.

Adams, Edward Dean. *Niagara Power: History of the Niagara Falls Power Company, 1886–1918*. Niagara Falls, NY: Niagara Falls Power Company, 1927.

Adams, Henry. *The Education of Henry Adams: An Autobiography*. Boston: Houghton Mifflin, 1918.

Annual of Scientific Discovery, 1869. Boston: Gould and Lincoln, 1869.

Annual Report of the American Railway Master Mechanics' Association. Cincinnati, OH: Wilstach, Baldwin, 1880.

Annual Report of the Board of Regents of the Smithsonian Institution. Washington, DC: Smithsonian Institution, 1901.

Ansolabehere, Stephen, and David M. Konisky. *Cheap and Clean: How Americans Think about Energy in the Age of Global Warming*. Cambridge: MIT Press, 2014.

Aughey, Samuel. *Sketches of the Physical Geography and Geology of Nebraska*. Omaha, NE: Daily Republican Book and Job Office, 1880.

"The Awakening of the Cable." *Littell's Living Age*, Oct.–Dec. 1866, 56–59.

Badger, Reid. *The Great American Fair: The World's Columbian Exposition and American Culture*. Chicago: Nelson Hall, 1979.

Ballard, Walter J. "Industrial and Commercial Progress since 1870." *Protectionist*, July 1902, 160.

Bancroft, Robert M., and Francis J. Bancroft. *Tall Chimney Construction: A Practical Treatise on the Construction of Tall Chimney Shafts Constructed in Brick, Stone, Iron and Concrete*. Manchester: J. Calvert, 1885.

Banner, Stuart. *The Death Penalty: An American History*. Cambridge, MA: Harvard University Press, 2003.

Baron, Robert, and Nicholas Spitzer. *Public Folklore*. Jackson: University Press of Mississippi, 2008.

Barrett, John Patrick. *Electricity at the Columbian Exposition: Including an Account of the Exhibits in the Electricity Building, the Power Plant in*

Machinery Hall, the Arc and Incandescent Lighting of the Grounds and Buildings. Chicago: R. R. Donnelley, 1894.

Barrows, Anna, et al. *Everyday Housekeeping: A Magazine for Practical Housekeepers and Mothers.* Boston: Clark-Clary, 1898.

Bartlett, Helen L. "The Efficiency of Electricity to the Housekeeper." *Electrical Review and Western Electrician* 64, no. 19 (June 1914): 934–35.

Basalla, George. *The Evolution of Technology.* Cambridge: Cambridge University Press, 1988.

Beecher, Catharine Esther, and Harriet Beecher Stowe. *The American Woman's Home: Or, Principles of Domestic Science; Being a Guide to the Formation and Maintenance of Economical, Healthful, Beautiful, and Christian Homes.* New York: J. B. Ford, 1869.

Bellamy, Edward. *Equality.* New York: Appleton, 1898.

Bellamy, Edward. *Looking Backward: 2000–1887.* Boston: Ticknor, 1888.

Bernard, Karl. *Travels through North America, during the Years 1825 and 1826.* Philadelphia: Carey, Lea and Carey, 1828.

Biello, David. "Where Did the Carter White House's Solar Panels Go?" *Scientific American,* August 6, 2010. Accessed October 8, 2015. http://www.scientificamerican.com/article/carter-white-house-solar-panel-array/.

Bigelow, Jacob. *Elements of Technology: Taken Chiefly from a Course of Lectures Delivered at Cambridge, on the Application of the Sciences to the Useful Arts.* Boston: Hilliard, Gray, Little and Wilkins, 1829.

Bijker, W. E., et al. *The Social Construction of Technological Systems: New Directions in the Sociology and History of Technology.* Cambridge, MA: MIT Press, 2012.

"Boiler House and Machine Shop." *Engineering.* London: Office for Advertisements and Publication, 1876.

Brimblecombe, Peter. *The Big Smoke: A History of Air Pollution in London since Medieval Times.* London: Routledge, 2011.

Brinley, Clarence C. "Reducing Cost with Mechanical Stokers." *Engineering Magazine,* November 1915, 276–92.

Britton, Diane F. *The Iron and Steel Industry in the Far West: Irondale, Washington.* Niwot: University Press of Colorado, 1991.

Brown, Dee Alexander. *The Year of the Century: 1876.* New York: Scribner, 1966.

Browne, Thomas. *The Works of Sir Thomas Browne.* London: Henry G. Bohn, 1852.

Bruce, Robert V. *Bell: Alexander Graham Bell and the Conquest of Solitude.* Ithaca, NY: Cornell University Press, 1990.

"The Brush Electric Light." *Scientific American* 44, no. 14 (April 2, 1881): 211.

Bryant, Edward. *Climate Process and Change*. Cambridge: Cambridge University Press, 1997.

Bull, Marcus. *Experiments to Determine the Comparative Value of the Principal Varieties of Fuel Used in the United States, and Also in Europe*. Philadelphia: J. Dobson, 1827.

Cao, Tian Yu. *Conceptual Developments of 20th Century Field Theories*. New York: Cambridge University Press, 1997.

Carlson, W. Bernard. *Innovation as a Social Process: Elihu Thomson and the Rise of General Electric*. Cambridge: Cambridge University Press, 2003.

Carlson, W. Bernard. *Tesla: Inventor of the Electrical Age*. Princeton, NJ: Princeton University Press, 2013.

Cashman, Sean Dennis. *America in the Gilded Age: From the Death of Lincoln to the Rise of Theodore Roosevelt*. New York: New York University Press, 1984.

Centennial Historical Discourses: Delivered in the City of Philadelphia, June, 1876. Philadelphia: Presbyterian Board of Publication, 1876.

Chandler, Alfred D., Jr. "Anthracite Coal and the Beginnings of the Industrial Revolution in the United States." *Business History Review* 46, no. 2 (1972): 141–81.

Chandler, Alfred D., Jr., and Takashi Hikino. *Scale and Scope: The Dynamics of Industrial Capitalism*. Cambridge, MA: Belknap Press of Harvard University Press, 1994.

Chandler, Charles Frederick. *Report on the Gas Nuisance in New York*. New York: D. Appleton, 1870.

Cheney, Margaret. *Tesla: Man out of Time*. New York: Simon and Schuster, 2001.

Christensen, Paul P. "Land Abundance and Cheap Horsepower in the Mechanization of the Antebellum United States Economy." *Explorations in Economic History* 18, no. 4 (1981): 309–29.

Church, William Conant. *The Life of John Ericsson*. Vol. 2. London: Sampson, Low, Marston, 1892.

Citizens' Association of Chicago, Committee on Smoke. *Report of Smoke Committee of the Citizens' Association of Chicago: May, 1889*. Chicago: G. E. Marshall, 1889.

"City of St. Louis v. Heitzeberg Packing & Provision Co." *Southwestern Reporter*, vol. 42, October 18, 1897–January 3, 1898, 954–57. Saint Paul, MN: West Publishing, 1898.

"Clean Electric Light." *Country Gentleman,* July 6, 1918, 25.

"Clean Electric Motors." *Popular Mechanics,* September 1, 1924, 122.

"Clean Electric Plants." *Motor Boating,* December 1, 1923, 98.

Cleave, Egbert. *Cleave's Biographical Cyclopaedia of the State of Ohio.* Cleveland, OH, 1875.

Cleveland, Cutler J. *Concise Encyclopedia of the History of Energy.* San Diego, CA: Elsevier, 2009.

Cohen, I. Bernard. *Benjamin Franklin's Science.* Cambridge, MA: Harvard University Press, 1990.

Collins, Francis Arnold. "Pops of Popular Science." *Boy's Life,* August 1920, 38.

Committee on Electricity in Economic Growth. *Electricity in Economic Growth.* Washington, DC: National Academy Press, 1986.

Cooke, William Fothergill. *The Electric Telegraph: Was It Invented by Professor Wheatstone?* London, 1854.

Copland, James. *A Dictionary of Practical Medicine.* Vol. 3. London: Longmans, Brown, Green, Longmans and Roberts, 1858.

Copland, James. *A Dictionary of Practical Medicine: Comprising General Pathology.* Vol. 1. New York: Harper and Brothers, 1845.

Coulson, Thomas. *Joseph Henry: His Life and Work.* Princeton, NJ: Princeton University Press, 1950.

Cowan, Ruth Schwartz. "The 'Industrial Revolution' in the Home: Household Technology and Social Change in the 20th Century." *Technology and Culture* 17, no. 1 (1976): 1–23.

Cowan, Ruth Schwartz. *More Work for Mother: The Ironies of Household Technology from the Open Hearth to the Microwave.* New York: Basic Books, 1983.

Coxe, Tench. *A View of the United States of America, in a Series of Papers, Written at Various Times, between the Years 1787 and 1794.* Philadelphia, 1794.

Coyle, Kevin. *Environmental Literacy in America: What Ten Years of NEETF/Roper Research and Related Studies Say About Environmental Literacy in the U.S.* Washington, DC: National Environmental Education and Training Foundation, 2005.

Crockett, D. *An Account of Col. Crockett's Tour to the North and Down East: In the Year of Our Lord One Thousand Eight Hundred and Thirty-Four. His Object Being to Examine the Grand Manufacturing Establishments of the Country; and Also to Find out the Condition of Its Literature and Its*

Morals, the Extent of Its Commerce, and the Practical Operation of "the Experiment." Philadelphia, 1835.

Cronon, William. *Changes in the Land: Indians, Colonists, and the Ecology of New England.* New York: Hill and Wang, 1983.

Cronon, William. *Nature's Metropolis: Chicago and the Great West.* New York: W. W. Norton, 1992.

Culvahouse, Tim. *The Tennessee Valley Authority: Design and Persuasion.* New York: Princeton Architectural Press, 2007.

Dalzell, Frank. *Engineering Invention: Frank J. Sprague and the U.S. Electrical Industry.* Cambridge, MA: MIT Press, 2010.

Daniels, George H., and Mark H. Rose, eds. *Energy and Transport: Historical Perspectives on Policy Issues.* Beverly Hills: Sage, 1982.

Darwin, Charles. *The Descent of Man and Selection in Relation to Sex.* New York: D. Appleton, 1872.

Dean, Teresa. *White City Chips.* Chicago: Warren, 1895.

Derry, Thomas Kingston, and Trevor I. Williams. *A Short History of Technology from the Earliest Times to A. D. 1900.* Oxford: Clarendon, 1960.

Devens, R. M. *American Progress: Or, the Great Events of the Greatest Century, Including Also Life Delineations of Our Most Noted Men. A Book for the Times.* Chicago: H. Heron, 1882.

Dickens, Charles. *American Notes for General Circulation.* London: Chapman and Hall, 1842.

Dickerson, E. N. *Joseph Henry and the Magnetic Telegraph.* New York: C. Scribner's Sons, 1885.

Di Cola, Joseph M., and David Stone. *Chicago's 1893 World's Fair.* Charleston, SC: Arcadia, 2012.

"Do You Glow with Health?" *Popular Mechanics,* March 1922, 177.

Dodd, Anna Bowman. *The Republic of the Future, or, Socialism a Reality.* New York: Cassell, 1887.

Donnelly, Ignatius. *Caesar's Column: A Story of the Twentieth Century.* Chicago: F. J. Schulte, 1890.

Dyer, Frank L. *Edison: His Life and Inventions.* New York: Harper and Brothers, 1910.

Edison Electric Light Company v. F. P. Little Electrical Construction and Supply Company et al: On Letters Patent No. 281,576. New York, 1896.

Edison, Thomas A. "The Dangers of Electric Lighting." *North American Review* 149 (November 1889): 625–35.

Edward, Shelly Charles. "Transactions of the Seventh International Congress of Hygiene and Demography." London, 1892.

Edwards, Paul N. "Infrastructure and Modernity: Force, Time, and Social Organization in the History of Sociotechnical Systems." In *Modernity and Technology,* edited by Thomas J. Misa, Philip Brey, and Andrew Feenberg, 185–226. Cambridge, MA: MIT Press, 2003.

Edwards, Rebecca. *New Spirits: Americans in the Gilded Age, 1865–1905.* New York: Oxford University Press, 2006.

"Effect of Strike upon Central-Station Supply to Coal Mines." *Electrical Review* 75 (November 29, 1919): 903–5.

"An Electrical Euclid." *Electric Power* 1, no. 9 (1889): 290.

"Electrical Magic." *Scientific American* 45, no. 21 (November 19, 1881): 327–29.

Emerson, Ralph Waldo. *The Collected Works of Ralph Waldo Emerson: Society and Solitude.* Edited by R. E. Spiller, A. R. Ferguson, J. Slater, and J. F. Carr. Cambridge, MA: Belknap Press of Harvard University Press, 1971.

Emerson, Ralph Waldo. "Conduct of Life." In *The Prose Works of Ralph Waldo Emerson: In Two Volumes.* Boston: J. R. Osgood, 1875.

Emerson, Ralph Waldo. *The Journals and Miscellaneous Notebooks of Ralph Waldo Emerson.* Edited by A. W. Plumstead. Cambridge, MA: Belknap Press of Harvard University Press, 1969.

Emerson, Ralph Waldo. *Nature.* Boston: J. Munroe, 1836.

Engs, Ruth Clifford. *Clean Living Movements: American Cycles of Health Reform.* Westport, CT: Praeger, 2000.

Ericsson, John. *Contributions to the Centennial Exhibition.* New York, 1876.

Etzler, John A. *The Paradise within the Reach of All Men, without Labor, by Powers of Nature and Machinery: An Address to All Intelligent Men.* Pittsburgh, PA: Etzler and Reinhold, 1833.

Etzler, John A. *Two Visions of J. A. Etzler . . . A Revelation of Futurity.* Surrey, 1844.

Evans, Lewis. *Geographical, Historical, Political, Philosophical and Mechanical Essays the First, Containing an Analysis of a General Map of the Middle British Colonies in America; and of the Country of the Confederate Indians; a Description of the Face of the Country; the Boundaries of the Confederates; and the Maritime and Inland Navigations of the Several Rivers and Lakes Contained Therein.* Philadelphia: Benjamin Franklin and D. Hall, 1755.

Evelyn, John. *Fumifugium, or, the Inconveniencie of the Aer and Smoak of London Dissipated Together with Some Remedies Humbly Proposed.* London, 1661.

"Extensions to the Plant of the Niagara Falls Power Company." *American Gas Light Journal,* January 17, 1898.

Farey, J. *A Treatise on the Steam-Engine, Historical, Practical and Descriptive.* London: Longman, 1827.

Fishback, Price V. *Soft Coal, Hard Choices: The Economic Welfare of Bituminous Coal Miners, 1890–1930.* New York: Oxford University Press, 1992.

Fisher, Alan C., Jr. "You and the Obedient Atom." *National Geographic,* vol. 114, no. 3, September 1958, 303–53.

"Fisk Street Station of the Commonwealth Edison Company, Chicago." *Western Electrician* 38, no. 1 (January 6, 1906): 1–6.

Fleming, James Rodger. *Historical Perspectives on Climate Change.* New York: Oxford University Press, 2005.

Francis, Charles S., Joseph H. Francis, and Alexander Anderson. *The Parlour Book.* Boston: Charles S. Francis, 1839.

Frank, Adam. "Climate Change Is Not Our Fault." *Cosmos & Culture.* NPR, October 28, 2015. Accessed November 10, 2015. http://www.npr.org/sections/13.7/2015/10/06/446109168/climate-change-is-not-our-fault.

Franklin, Benjamin. *Experiments and Observations on Electricity, Made at Philadelphia in America.* London, 1769.

Freese, Barbara. *Coal: A Human History.* Cambridge, MA: Perseus, 2003.

Friedman, Lawrence Jacob, and Mark Douglas McGarvie. *Charity, Philanthropy, and Civility in American History.* Cambridge: Cambridge University Press, 2003.

Fuel Magazine: The Coal Operators National Weekly. Fuel Publishing, 1909.

Galloway, Elijah Hebert Luke. *History and Progress of the Steam Engine: With a Practical Investigation of Its Structure and Application.* London: T. Kelly, 1836.

"The Gaulard-Gibbs Secondary Generator." Supplement, *Scientific American* 15, no. S387 (June 2, 1883): 6172.

"Generation of Electricity in the Largest Steam-Driven Station." *Electrical Review and Western Electrician* 57, no. 23 (1910): 1124.

Gibson, Charles Robert. *The Romance of Modern Electricity: Describing in Non-Technical Language, What Is Known about Electricity and Many of Its Interesting Applications.* Philadelphia: J. B. Lippincott, 1906.

Gillette, King Camp. *The Human Drift.* Boston: New Era, 1894.

Gladden, Washington. *Working People and Their Employers.* New York: Funk, 1888.

Gonzalez, George A. *The Politics of Air Pollution: Urban Growth, Ecological Modernization, and Symbolic Inclusion.* Albany: State University of New York Press, 2005.

Goss, W. *Smoke Abatement and Electrification of Railway Terminals in Chicago.* Report of the Chicago Association of Commerce, Committee of Investigation on Smoke Abatement and Electrification of Railway Terminals. Chicago: Chicago Association of Commerce, Committee of Investigation on Smoke Abatement Industry, 1915.

Grauvogl, Eduard von, and Geo E. Shipman. *Text Book of Homoeopathy.* Chicago: C. S. Halsey, 1870.

"The Great Exhibition." *New York Times,* June 12, 1876.

Green, Mary Anne Everett, et al. *Calendar of State Papers, Domestic Series, of the Reign of Charles II, 1660–1685.* London: H. M. Stationery Office, 1860.

Griscom, John H. *The Uses and Abuses of Air: Showing Its Influence in Sustaining Life, and Producing Disease; with Remarks on the Ventilation of Houses, and the Best Methods of Securing a Pure and Wholesome Atmosphere inside of Dwellings, Churches, Courtrooms, Workshops, and Buildings of All Kinds.* New York: Redfield, 1848.

Gross, Linda P., and Theresa R. Snyder. *Philadelphia's 1876 Centennial Exhibition.* Charleston, SC: Arcadia, 2005.

Gugliotta, Angela. "Class, Gender, and Coal Smoke: Gender Ideology and Environmental Injustice in Pittsburgh, 1868–1914." *Environmental History* 5, no. 2 (2000): 165–93.

Gurevich, Vladimir. *Electric Relays: Principles and Applications.* Boca Raton, FL: Taylor and Francis, 2005.

Hale, Nathan. *Chronicle of Events, Discoveries, and Improvements, for the Popular Diffusion of Useful Knowledge, with an Authentic Record of Facts. Illustrated with Maps and Drawings.* Boston: S. N. Dickinson, 1850.

Hammond, John Craig. "New Business for Electric Central Stations." *Cassier's Magazine,* vol. 30, no. 3, July 1906, 256–60.

Hankins, Thomas L. *Science and the Enlightenment.* Cambridge: Cambridge University Press, 1985.

Harman, Peter Michael. *Energy, Force, and Matter: The Conceptual Development of Nineteenth-Century Physics.* Cambridge: Cambridge University Press, 1995.

Harriot, Thomas. *A Briefe and True Report of the New Found Land of Virginia*. New York: Dover, 1972. First published 1590.

Hart, Joseph K. "Neighborhood and Community Life." Paper presented at the National Conference of Social Work, Cleveland, OH, 1926. Accessed October 8, 2014, http://quod.lib.umich.edu/n/ncosw/ACH8650.1926.001?rgn=main;view=fulltext.

Hausman, William J., Peter Hertner, and Mira Wilkins. *Global Electrification: Multinational Enterprise and International Finance in the History of Light and Power, 1878–2007*. New York: Cambridge University Press, 2008.

Hays, Samuel P. *The Response to Industrialism, 1885–1914*. Chicago: University of Chicago Press, 1995.

Hays, Samuel P., and Barbara D. Hays. *Beauty, Health, and Permanence: Environmental Politics in the United States, 1955–1985*. Cambridge: Cambridge University Press, 1987.

Hazard, Samuel. *Hazard's United States Commercial and Statistical Register*. Philadelphia: W. F. Geddes, 1842.

Hazard, Samuel. *The Register of Pennsylvania: Devoted to the Preservation of Facts and Documents and Every Other Kind of Useful Information Respecting the State of Pennsylvania*. Philadelphia: W. F. Geddes, 1829.

Heertje, Arnold, and Mark Perlman. *Evolving Technology and Market Structure: Studies in Schumpeterian Economics*. Ann Arbor: University of Michigan Press, 1990.

Heilbron, John L. *Electricity in the 17th and 18th Centuries: A Study of Early Modern Physics*. Berkeley: University of California Press, 1979.

Heilbron, John L. *The Oxford Companion to the History of Modern Science*. Oxford: Oxford University Press, 2003.

Hennepin, Louis, Louis Joliet, and Jacques Marquette. *A New Discovery of a Vast Country in America Extending above Four Thousand Miles, between New France and New Mexico. With a Description of the Great Lakes, Cataracts, Rivers, Plants, and Animals*. London, 1698.

Henry, Joseph. *A Memorial of Joseph Henry*. Washington, DC: US Government Printing Office, 1880.

Henry, Joseph. "On a Reciprocating Motion Produced by Magnetic Attraction and Repulsion." *American Journal of Science and Arts* 20, no. 2 (July 1831): 340–43.

Henry, Joseph. *Scientific Writings of Joseph Henry*. Washington, DC: Smithsonian Institution, 1886.

Henry, Mary A. "The Electro-Magnet; or Joseph Henry's Place in the History of the Electro Magnetic Telegraph." *Electrical Engineer* 17, no. 297 (Jan. 10, 1894): 26–28.

Higginson, Francis. *New Englands Plantation; or, a Short and True Description of the Commodities and Discommodities of That Country.* London, 1860.

Hingston, Edward Peron. *The Genial Showman. Being Reminiscences of the Life of Artemus Ward; and Pictures of a Showman's Career in the Western World.* London: J. C. Hotten, 1871.

Hirsh, Richard F., and Benjamin K. Sovacool. "Wind Turbines and Invisible Technology: Unarticulated Reasons for Local Opposition to Wind Energy." *Technology and Culture* 54, no. 4 (2013): 705–34.

Hirtenstein, Anna. "Makai Builds Ocean Thermal-Energy Demo Plant in Hawaii." *Bloomberg Business News,* August 21, 2015. Accessed October 8, 2015. http://www.bloomberg.com/news/articles/2015-08-21/makai-builds-ocean-thermal-energy-demo-plant-in-hawaii.

Hochfelder, David. *The Telegraph in America, 1832–1920.* Baltimore, MD: Johns Hopkins University Press, 2012.

Hofstadter, Richard. *The Age of Reform: From Bryan to F.D.R.* New York: Vintage Books, 1955.

Hopkins, Samuel, and Edwards Park. *The Works of Samuel Hopkins: With a Memoir of His Life and Character.* Boston: Doctrinal Tract and Book Society, 1854.

Hounshell, David A. *From the American System to Mass Production, 1800–1932: The Development of Manufacturing Technology in the United States.* Baltimore, MD: Johns Hopkins University Press, 1985.

"How to Make Pleasant Homes." *Godey's Magazine,* vol. 92, March 1876, 280–81.

Howells, William Dean. "A Sennight at the Centennial." *Atlantic,* July 1876.

Howells, William Dean. *A Traveler from Altruria: Romance.* New York: Harper and Brothers, 1894.

Hughes, Thomas Parke. *Networks of Power: Electrification in Western Society, 1880–1930.* Baltimore, MD: Johns Hopkins University Press, 1983.

"Hydro Electric Development." *Power* 47, no. 2 (June 30, 1918): 236–37.

Ide, George R. "Smoke Abatement in Cities." *Proceedings of the Engineers' Club of Philadelphia* 9, no. 2 (1892): 141–52.

Iles, George. "Why Progress Is by Leaps." *Popular Science,* June 1896, 216–30.

Ingram, J. S. *The Centennial Exposition Described and Illustrated: Being a*

Concise and Graphic Description of This Grand Enterprise Commemorative of the First Centennary [sic] of American Independence. Philadelphia: Hubbard Bros., 1876.

Insull, Samuel. *Central-Station Electric Service; Its Commercial Development and Economic Significance as Set Forth in the Public Addresses (1897–1914) of Samuel Insull*. Chicago: Privately printed, 1915.

Insull, Samuel. "The Progress of Economic Power Generation and Distribution." *Journal of the American Society of Mechanical Engineers* 38 (1916): 845–54.

Jackson, Ben, Marc Stears, and Michael Freeden, eds. *Liberalism as Ideology: Essays in Honour of Michael Freeden*. Oxford: Oxford University Press, 2012.

Jacobson, Mark Z. *Atmospheric Pollution: History, Science, and Regulation*. Cambridge: Cambridge University Press, 2002.

Jacoby, George W., and Ralph J. Jacoby. *Electricity in Medicine; a Practical Exposition of the Methods and Use of Electricity in the Treatment of Disease, Comprising Electrophysics, Apparatus, Electrophysiology and Electropathology, Electrodiagnosis and Electroprognosis, Generalelectrotherapeutics and Special Electrotherapeutics*. Philadelphia: P. Blakiston's Son, 1919.

Jefferson, Thomas. *Memoirs, Correspondence and Private Papers of Thomas Jefferson, Late President of the United States*. Edited by Thomas Jefferson Randolph. London: Colburn and Bentley, 1829.

Jefferson, Thomas. *Notes on the State of Virginia*. London: John Stockwell, 1787.

Jefferson, Thomas. *The Writings of Thomas Jefferson: Being His Autobiography, Correspondence, Reports, Messages, Addresses, and Other Writings, Official and Private: Published by the Order of the Joint Committee of Congress on the Library, from the Original Manuscripts, Deposited in the Department of State*. Vol. 5. Washington, DC: Taylor and Maury, 1854.

Jefferson, Thomas, et al. *The Writings of Thomas Jefferson: Containing His Autobiography, Notes on Virginia, Parliamentary Manual, Official Papers, Messages and Addresses, and Other Writings, Official and Private, Now Collected and Published in Their Entirety for the First Time*. Washington, DC: Thomas Jefferson Memorial Association, 1904.

"John Ericsson." *Science* 13, no. 319 (Mar. 15, 1889): 189–91.

John, Richard R. *Network Nation: Inventing American Telecommunications*. Cambridge, MA: Belknap Press of Harvard University Press, 2010.

Johnson, Rossiter. *A History of the World's Columbian Exposition Held in Chicago in 1893.* New York: D. Appleton, 1897.

Jones, Christopher F. *Routes of Power: Energy and Modern America.* Cambridge, MA: Harvard University Press, 2014.

Jonnes, Jill. *Empires of Light: Edison, Tesla, Westinghouse, and the Race to Electrify the World.* New York: Random House, 2003.

Josephson, Matthew. *Edison: A Biography.* New York: McGraw-Hill, 1959.

Kasson, John F. *Civilizing the Machine: Technology and Republican Values in America, 1776–1900.* New York: Grossman, 1976.

Keats, John. *Selected Letters of John Keats.* Edited and with an introduction by Lionel Trilling. New York: Farrar, Straus and Young, 1951.

Kelly, Walt. *Pogo: We Have Met the Enemy and He Is Us.* New York: Simon and Schuster, 1972.

Klein, Maury. *The Power Makers: Steam, Electricity, and the Men Who Invented Modern America.* 1st US ed. New York: Macmillan, 2008.

Krause, W. H. *Ericsson Cycle Gas Turbine Powerplants.* Santa Monica, CA: RAND Corporation, 1979. Accessed October 8, 2015. http://www.rand.org/pubs/reports/R2327.

Lane, Carter N. *Acid Rain: Overview and Abstracts.* New York: Nova Science, 2003.

Lane, Mary E. Bradley. *Mizora: A World of Women.* Lincoln: University of Nebraska Press, 1999.

Lasch, Christopher. *The Minimal Self: Psychic Survival in Troubled Times.* New York: W. W. Norton, 1984.

Leach, William R. *Land of Desire: Merchants, Power, and the Rise of a New American Culture.* New York: Knopf Doubleday, 1993.

Leading Manufacturers and Merchants of Cincinnati and Environs: The Great Railroad Centre of the South and Southwest. New York: International Publishing, 1886.

Lears, T. J. Jackson. *No Place of Grace: Antimodernism and the Transformation of American Culture, 1880–1920.* Chicago: University of Chicago Press, 1994.

Licht, Walter. *Industrializing America: The Nineteenth Century.* Baltimore, MD: Johns Hopkins University Press, 1995.

Life and Light for Heathen Women. Vol. 6. Boston: Rand, Avery, 1876.

Lincoln, Abraham. *Abraham Lincoln: Complete Works, Comprising His Speeches, Letters, State Papers, and Miscellaneous Writings.* Edited by J. G. Nicolay and J. Hay. New York: Century, 1894.

Lincoln, Abraham. *Collected Works of Abraham Lincoln.* Edited by Roy P. Basler. Vol. 2, *1848–1858.* New Brunswick, NJ: Rutgers University Press, 1953.

Lincoln, Paul M. "Modern Developments in Power Generation." *Electrical Review and Western Electrician* 64, no. 24 (June 27, 1914): 1301–2.

Linebaugh, Peter. *The London Hanged: Crime and Civil Society in the Eighteenth Century.* Cambridge: Cambridge University Press, 1992.

Linebaugh, Peter. *The Magna Carta Manifesto: Liberties and Commons for All.* Berkeley: University of California Press, 2008.

Loy, Daniel Oscar. *Poems of the White City.* Chicago: D. O. Loy, 1893.

Mackay, Alexander. *The Western World: Or, Travels in the United States in 1846–47: Exhibiting Them in Their Latest Development, Social, Political, and Industrial.* London: R. Bentley, 1849.

MacMillan, Donald. *Smoke Wars: Anaconda Copper, Montana Air Pollution, and the Courts, 1890–1924.* Helena: Montana Historical Society Press, 2000.

Maddocks, Mildred. "We Recommend Electricity." *Good Housekeeping,* June 1918, 57, 139.

Mantoux, Paul. *The Industrial Revolution in the Eighteenth Century: An Outline of the Beginnings of the Modern Factory System in England.* London: Taylor and Francis, 1928.

Martin, Thomas C. *Forty Years of Edison Service, 1882–1922: Outlining the Growth and Development of the Edison System in New York City.* New York: New York Edison Company, 1922.

Martineau, Harriet. *Society in America.* London: Saunders and Otley, 1837.

Marvin, Carolyn. *When Old Technologies Were New: Thinking about Electric Communication in the Late Nineteenth Century.* New York: Oxford University Press, 1988.

Marx, Karl, and Frederick Engels. *Capital, Volume One: A Critique of Political Economy.* London: Dover Publications, 2012.

Marx, Leo. *The Machine in the Garden: Technology and the Pastoral Ideal in America.* New York: Oxford University Press, 1964.

Maslin, Mark. *Global Warming: A Very Short Introduction.* Oxford: Oxford University Press, 1979.

Massachusetts Medical Society. *Medical Communications.* Vol. 8. Boston: Massachusetts Medical Society, 1854.

Matas, R. R. A. *Bielas Y Álabes 1826–1914.* Ministerio de Industria, Turismo y Comercio. Oficina Española de Patentes y Marcas, 2008.

Mather, W. W. *First Annual Report on the Geological Survey of the State of Ohio*. Columbus, OH: S. Medary, 1838.

Mathews, J. A. *Report upon Smoke Abatement: An Impartial Investigation of the Ways and Means of Abating Smoke, Results Attained in Other Cities, Merits of Patented Devices, Together with Practical Suggestions to the Department of Smoke Abatement, the Steam Plant Owner and the Private Citizen*. Syracuse, NY: Syracuse Chamber of Commerce, 1907.

McCabe, James D. *The Illustrated History of the Centennial Exhibition: Held in Commemoration of the One Hundreth Anniversary of American Independence*. Philadelphia: National Publishing Company, 1876.

McComas, Alan J. *Galvani's Spark: The Story of the Nerve Impulse*. New York: Oxford University Press, 2011.

McDonald, Forrest. *Insull*. Chicago: University of Chicago Press, 1962.

McGreevy, Patrick Vincent. *Imagining Niagara: The Meaning and Making of Niagara Falls*. Amherst: University of Massachusetts Press, 1994.

McNeill, John R. *Something New under the Sun: An Environmental History of the Twentieth-Century World*. New York: W. W. Norton, 2000.

McNerney, James, Jessika E. Trancik, and J. Doyne Farmer, "Historical Costs of Coal-Fired Electricity and Implications for the Future." Santa Fe Institute Working Paper 2009–12–047. Santa Fe, NM, December 16, 2009.

Melosi, Martin V. *Effluent America: Cities, Industry, Energy, and the Environment*. Pittsburgh, PA: University of Pittsburgh Press, 2001.

Melosi, Martin V. *The Sanitary City: Environmental Services in Urban America from Colonial Times to the Present*. Pittsburgh, PA: University of Pittsburgh Press, 2008.

Melville, Herman. *Moby-Dick: Or, the Whale*. New York: Harper and Brothers, 1851.

Merriam, J. C. *Eighty Years Progress of the United States, Showing the Various Channels of Industry and Education through Which the People of the United States Have Arisen from a British Colony to Their Present National Importance*. Hartford, CT: L. Stebbins, 1865.

Miller, Perry. "The Responsibility of Mind in a Civilization of Machines." *American Scholar* 31, no. 1 (1961): 51–69.

Modern Machinery. Chicago: Modern Machinery Publishing Company, 1905.

Moffit, Cleveland. "Stories from the Archives of the Pinkerton Detective

Agency." *McClure's Magazine*, 1894.

Mollenhoff, David V. *Madison, a History of the Formative Years.* Madison: University of Wisconsin Press, 2003.

Monell, Samuel H. *Electricity in Health and Disease; a Treatise of Authentic Facts for General Readers, in Which Is Shown How Electric Currents Are Made to Act as Curative Remedies, Together with an Account of the Principal Diseases Which Are Benefited by Them.* New York: McGraw, 1907.

Moody, John. *Moody's Manual of Investments: American and Foreign.* New York: Moody's Investors Service, 1922.

Morris, Israel W. *The Duty on Coal; Being a Few Facts Connected with the Coal Question, Which Will Furnish Matter for Thought to the Friends of American Industry.* Philadelphia: Baird, 1872.

Morse, Edward L., ed. *Samuel F. B. Morse: His Letters and Journals.* Boston: Houghton Mifflin, 1914.

Morus, Iwan Rhys. *When Physics Became King.* Chicago: University of Chicago Press, 2005.

Moyer, Albert E. *Joseph Henry: The Rise of an American Scientist.* Washington, DC: Smithsonian Institution Press, 1997.

"Mr. Brush's Windmill Dynamo." *Scientific American* 63, no. 25 (December 20, 1890): 389.

Mueller, H. F. "The 'One Hundredth Anniversary' of George Henry Corliss." *Power,* May 14, 1918, 682–87.

Muir, Diana. *Reflections in Bullough's Pond: Economy and Ecosystem in New England.* Lebanon, NH: University Press of New England, 2000.

Mumford, Lewis. *The City in History: Its Origins, Its Transformations, and Its Prospects.* New York: Harcourt, Brace and World, 1961.

Mumford, Lewis. *Technics and Civilization.* 1934. Reprint, Chicago: University of Chicago Press, 2010.

Myers, David M. *The Power Plant.* New York: Industrial Extension Institute, 1922.

Nash, Roderick Frazier. *Wilderness and the American Mind.* New Haven, CT: Yale University Press, 2001.

Nelson, Daniel. *Farm and Factory: Workers in the Midwest, 1880–1990.* Bloomington: Indiana University Press, 1995.

Neufeldt, Leonard N. "The Science of Power: Emerson's Views on Science and Technology in America." *Journal of the History of Ideas* 38, no. 2 (1977): 329–44.

"New Outlook." *New Outlook Magazine,* July 27, 1895, 128.

Newlin, Keith. *Hamlin Garland: A Life.* Lincoln: University of Nebraska Press, 2008.

Noble, David F. *America by Design: Science, Technology, and the Rise of Corporate Capitalism.* Oxford: Oxford University Press, 1979.

Noble, David F. *The Religion of Technology: The Divinity of Man and the Spirit of Invention.* New York: A. A. Knopf, 1997.

Northrop, Henry Davenport. *The World's Fair as Seen in One Hundred Days.* Philadelphia: National Publishing, 1893.

Nye, David E. *America as Second Creation: Technology and Narratives of New Beginnings.* Cambridge: MIT Press, 2003.

Nye, David E. *American Technological Sublime.* Cambridge: MIT Press, 1994.

Nye, David E. *Consuming Power: A Social History of American Energies.* Cambridge: MIT Press, 2001.

Nye, David E. *Electrifying America: Social Meanings of a New Technology, 1880–1940.* Cambridge: MIT Press, 1990.

"Ohio Census Records Meigs—Morgan County Census Pre-1830 Deeds Index." Accessed July 6, 2014. http://www.censusfinder.com/ohio-census-records9.htm.

Olenick, Richard P., Tom M. Apostol, and David L. Goodstein. *Beyond the Mechanical Universe: From Electricity to Modern Physics.* Cambridge: Cambridge University Press, 1986.

Olenick, Richard P., Tom M. Apostol, and David L. Goodstein. *The Mechanical Universe: Introduction to Mechanics and Heat.* Cambridge: Cambridge University Press, 2008.

Olsson, Mats-Olov, and Gunnar Sjöstedt. *Systems Approaches and Their Application: Examples from Sweden.* Dordrecht: Springer, 2004.

One Hundred Years' Progress of the United States. Hartford, CT: L. Stebbins, 1870.

Orrkok, Robert H., and George A. Fernald. *Engineering of Power Plants.* New York: McGraw-Hill, 1921.

Pamphlets on Insurance. New York: J. H. and C. M. Goodsell, 1871.

Park, Chris C. *Acid Rain: Rhetoric and Reality.* New York: Routledge, 2013.

Patterson, Walt. *Keeping the Lights On: Towards Sustainable Electricity.* Baltimore, MD: Brookings Institution Press, 2012.

Pennsylvania and the Centennial Exposition: Comprising the Preliminary and Final Reports of the Pennsylvania Board of Centennial Managers,

Made to the Legislature at the Sessions of 1877–8. Philadelphia: Gillin and Nagle, 1878.

Pera, Marcello, and Jonathan Mandelbaum. *The Ambiguous Frog: The Galvani-Volta Controversy on Animal Electricity.* Princeton, NJ: Princeton University Press, 1992.

Perlin, John. *A Forest Journey: The Role of Wood in the Development of Civilization.* 1st ed. New York: W. W. Norton, 1989.

Pfaelzer, Jean. *The Utopian Novel in America, 1886–1896: The Politics of Form.* Pittsburgh, PA: University of Pittsburgh Press, 1985.

Pimentel, David. "Ethanol Fuels: Energy Balance, Economics, and Environmental Impacts Are Negative." *Natural Resources Research* 12, no. 2 (2003): 127–34.

Platt, Harold L. *The Electric City: Energy and the Growth of the Chicago Area, 1880–1930.* Chicago: University of Chicago Press, 1991.

Pliny the Elder. *The Natural History.* Vol. 1. London: George Bell and Sons, 1893.

Podobnik, Bruce. *Global Energy Shifts: Fostering Sustainability in a Turbulent Age.* Philadelphia: Temple University Press, 2008.

Porter, George Richardson. *Useful Arts: A Treatise on the Origin, Progressive Improvement, and Present State of the Silk Manufacture.* Philadelphia: Carey and Lea, 1832.

Post, Rev. T. M. "The Outlook of the Times in Reference to the Progress of Christianity."*Missionary Herald* 78, no. 1, January 1882, 20.

Prasad, R. *Fundamentals of Electrical Engineering.* New Delhi: Prentice Hall, 2005.

Prime, Samuel I. *The Life of Samuel F. B. Morse, LL. D.: Inventor of the Electro-Magnetic Recording Telegraph.* New York: D. Appleton, 1875.

"Proceedings of the National Conference of Social Work Annual Session Held in Cleveland." Paper presented at the National Conference of Social Work, Cleveland, OH, 1926.

Proceedings of the One Hundredth Anniversary of the Introduction and Adoption of the "Resolutions Respecting Independency." Held in Philadelphia on the Evening of June 7, 1876, at the Pennsylvania Academy of the Fine Arts, and on July 1, 1876, at the Hall of Independence. Philadelphia: Collins, 1876.

Prout, Henry G. *A Life of George Westinghouse.* New York: C. Scribner, 1922.

"The Public Be———." *Chicago Journal of Commerce and Metal Industries* 61, no. 16, October 30, 1892, 22.

Purchas, Samuel. *Purchas His Pilgrimes: In Five Bookes.* London, 1625.
Pursell, Carroll W. *Technology in America: A History of Individuals and Ideas.* 2nd ed. Cambridge: MIT Press, 1990.
Pyne, Stephen J. *Fire: A Brief History.* Seattle: University of Washington Press, 2001.
Randall, D. T., and H. W. Weeks. *The Smokeless Combustion of Coal in Boiler Plants: With a Chapter on Central Heating Plants.* Vol. 2. Washington, DC: US Government Printing Office, 1909.
Rashed, Roshdi, and Régis Morelon, eds. *Encyclopedia of the History of Arabic Science: Technology, Alchemy and Life Sciences.* London: Routledge, 1996.
Reese, John S. *Guide Book for the Tourist and Traveler over the Valley Railway: The Short Line between Cleveland, Akron, and Canton.* Kent, OH: Kent State University Press in cooperation with Cuyahoga Valley National Park and Cuyahoga Valley Scenic Railroad, 2002.
"Report of National Commodity Advertisers' Division Advertising and Publicity Service Bureau." *National Electric Light Association Bulletin* 7, no. 7 1920, 553–55.
Report of Smoke Committee of the Citizens' Association of Chicago: May, 1889. Chicago: G. E. Marshall, 1889.
Righter, Robert W. *Wind Energy in America: A History.* Norman: University of Oklahoma Press, 1996.
Riis, Jacob A. *Children of the Tenements.* New York: Macmillan, 1904.
Robertson, Linda L. *Wabash County History Bicentennial Edition 1976, Wabash, Indiana.* Marceline, MO: Walsworth, 1976.
Roebroeks, Wil, Paola Villa, and Erik Trinkaus. "On the Earliest Evidence for Habitual Use of Fire in Europe." *Proceedings of the National Academy of Sciences of the United States of America* 108, no. 13 (2011): 5209–14.
Roemer, Kenneth M. *The Obsolete Necessity: America in Utopian Writings, 1888–1900.* Kent, OH: Kent State University Press, 1976.
Roosevelt, Franklin D. "Campaign Address in Portland, Oregon on Public Utilities and Development of Hydro-Electric Power." September 21, 1932. Online by Gerhard Peters and John T. Woolley, *The American Presidency Project.* Accessed October 8, 2015. http://www.presidency.ucsb.edu/ws/?pid=88390.
Rosen, Christine Meisner. "Businessmen against Pollution in Late Nineteenth Century Chicago." *Business History Review* 69, no. 3 (1995): 351–97.

Rosen, William. *The Most Powerful Idea in the World: A Story of Steam, Industry, and Invention.* Chicago: University of Chicago Press, 2012.

Rosenberg, Nathan, and Manuel Trajtenberg. "A General Purpose Technology at Work: The Corliss Steam Engine in the Late 19th Century U.S." National Bureau of Economic Research Working Paper Series No. 8485. Cambridge, MA: National Bureau of Economic Research, 2001.

Rossiter, Paul L. *The Electrical Resistivity of Metals and Alloys.* New York: Cambridge University Press, 1987.

Roy, Andrew. *The Coal Mines: Containing a Description of the Various Systems of Working and Ventilating Mines, Together with a Sketch of the Principal Coal Regions of the Globe, Including Statistics of the Coal Production.* Cleveland, OH: Robison, Savage, 1876.

Rutherford, Janice Williams. *Selling Mrs. Consumer: Christine Frederick and the Rise of Household Efficiency.* Athens: University of Georgia Press, 2010.

Rydell, Robert W. *All the World's a Fair: Visions of Empire at American International Expositions, 1876–1916.* Chicago: University of Chicago Press, 2013.

Sabatier, Paul. "An Advocacy Coalition Framework of Policy Change and the Role of Policy-Oriented Learning Therein." *Policy Sciences* 21, no. 2–3 (1988): 129–68.

Saltonstall, Leverett. *Report of the Massachusetts State Commissioner to the Centennial Exhibition at Philadelphia.* Boston: A. J. Wright, 1877.

Saward, Frederick Edward. "Saward's Coal Freight Circular." New York: F. E. Saward, 1869.

Schiffer, Michael B. *Power Struggles: Scientific Authority and the Creation of Practical Electricity before Edison.* Cambridge: MIT Press, 2008.

Schiffer, Michael B., Kacy L. Hollenback, and Carrie L. Bell. *Draw the Lightning Down: Benjamin Franklin and Electrical Technology in the Age of Enlightenment.* Berkeley: University of California Press, 2003.

Schurr, Sam H., and Bruce Carlton Netschert. *Energy in the American Economy, 1850–1975; an Economic Study of Its History and Prospects.* Baltimore, MD: Johns Hopkins University Press, 1960.

Sellers, Charles Grier. *The Market Revolution: Jacksonian America, 1815–1846.* New York: Oxford University Press, 1991.

"The Sewage Question: The Relation of Town and Country." *Farmers Magazine,* November 1859.

Shanahan, Timothy. "Kant, Naturphilosophie, and Oersted's Discovery of

Electromagnetism: A Reassessment." *Studies in History and Philosophy of Science* 20, no. 3 (1989): 287–305.

Shearer, D. A. "Coal-Pit Mine Mouth Power Plants." *Power* 47, no. 2, June 30, 1918.

Sloan, John. "Centennial Galop." Philadelphia: Lee and Walker, 1876.

Smil, Vaclav. *Energy in World History*. Boulder, CO: Westview Press, 1994.

Smil, Vaclav. *Energy Transitions: History, Requirements, Prospects*. Santa Barbara, CA: Praeger, 2010.

Smith, E. B., and M. L. Newell. *Reports of Cases Decided in the Appellate Courts of the State of Illinois*. Chicago: Callaghan, 1893.

Smith, Henry Nash. *Virgin Land: The American West as Symbol and Myth*. Cambridge, MA: Harvard University Press, 1950.

Smith, John, et al., *The Generall Historie of Virginia, New-England, and the Summer Isles: With the Names of the Adventurers, Planters, and Governours from Their First Beginning An: 1584 to This Present 1626*. London, 1632.

Smith, Merritt Roe, and Leo Marx, eds. *Does Technology Drive History? The Dilemma of Technological Determinism*. Cambridge: MIT Press, 1994.

Smith, Susan Harris. *American Drama: The Bastard Art*. Cambridge: Cambridge University Press, 2006.

Smoke Investigations: Bulletin, No. 1–10: 1912–1922. Pittsburgh, PA: Mellon Institute of Industrial Research, 1922.

"The Smoke Nuisance and Its Regulation, with Special Reference to the Condition Prevailing in Philadelphia." *Journal of the Franklin Institute* 144, July 1897, 17–61.

"Smoke Prevention: Report of the Special Committee on Prevention of Smoke." *Journal of the Association of Engineering Societies* 11 (1892): 291–327.

Somavat, Pavel, and Vinod Namboodiri. "Energy Consumption of Personal Computing Including Portable Communication Devices." *Journal of Green Engineering*, July 2012.

Southwell, B. G., J. J. Murphy, J. E. DeWaters, and P. A. LeBaron. *Americans' Perceived and Actual Understanding of Energy*. RTI Press publication No. RR-0018-1208. Research Triangle Park, NC: RTI Press, 2012. Accessed November 15, 2015, http://www.rti.org/rtipress.

Sovacool, Benjamin K. *Contesting the Future of Nuclear Power: A Critical Global Assessment of Atomic Energy*. Hackensack, NJ: World Scientific, 2011.

Sperling, Daniel, and Deborah Gordon. *Two Billion Cars: Driving toward Sustainability.* Foreword by Governor Arnold Schwarzenegger. New York: Oxford University Press, 2009.

"Standards Activities." Last modified November 3, 2015. Accessed May 3, 2014. http://www.ieeeghn.org/wiki/index.php/AIEE_History_1884-1963.

Steinhart, Eric Charles. *The Logic of Metaphor: Analogous Parts of Possible Worlds.* Dordrecht, Netherlands: Kluwer Academic Publishers, 2001.

Stokes, Melvyn, and Stephen Conway, eds. *The Market Revolution in America: Social, Political, and Religious Expressions, 1800-1880.* Charlottesville: University Press of Virginia, 1996.

Stoll, Steven. *The Great Delusion: A Mad Inventor, Death in the Tropics, and the Utopian Origins of Economic Growth.* New York: Hill and Wang, 2008.

Stone, Jacob L. *A Collection of Thoughts: Or, Key to Scripture. An Explanation of the Old and New Testaments, According to Reason, Nature and Existing Facts.* Chicago, 1881.

Stradling, David. *Conservation in the Progressive Era: Classic Texts.* Seattle: University of Washington Press, 2004.

Stradling, David. *Smokestacks and Progressives: Environmentalists, Engineers and Air Quality in America, 1881-1951.* Baltimore, MD: Johns Hopkins University Press, 1999.

Stradling, David, and Peter Thorsheim. "The Smoke of Great Cities: British and American Efforts to Control Air Pollution, 1860-1914." *Environmental History* 4, no. 1 (1999): 6-31.

Strong, Josiah. *Our Country: Its Possible Future and Its Present Crisis.* New York: Baker and Taylor, 1885.

Sullivan, Joseph P. "Fearing Electricity: Overhead Wire Panic in New York City." *IEEE Technology and Society Magazine* 14, no. 3 (1995): 8-16.

Sussex Archaeological Society. "Sussex Archaeological Collections Relating to the History and Antiquities of the County." Lewes, England: Sussex Archaeological Society, 1848.

Tariff Acts Passed by the Congress of the United States from 1789 to 1895: Including All Acts, Resolutions, and Proclamations Modifying or Changing Those Acts. Washington, DC: US Government Printing Office, 1896.

Tarkington, Booth. *The Turmoil: A Novel.* New York: Harper and Brothers, 1915.

Taylor, William B. *A Memoir of Joseph Henry: A Sketch of His Scientific Work.* Philadelphia: Collins, 1879.

Temin, Peter. *Iron and Steel in Nineteenth-Century America, an Economic Inquiry.* Cambridge: MIT Press, 1964.

Tesla, Nikola. "Our Future Motive Power." *Everyday Science and Mechanics,* December 1931, 230–36.

Tesla, Nikola. "Tesla's Speech: The Age of Electricity." *Cassier's Magazine,* March 1897, 381.

Thirsk, Joan. *Chapters from the Agrarian History of England and Wales, 1500–1750.* New York: Cambridge University Press, 1990.

Thomas de la Peña, Carolyn. *The Body Electric.* New York: NYU Press, 2005.

Thompson, S. P. *Dynamo-Electric Machinery: A Manual for Students of Electrotechnics.* London: E. and F. N. Spon, 1888.

Thompson, Silvanus P. *The Life of Lord Kelvin.* Providence, RI: American Mathematical Society, 2005. First published 1976 by Chelsea Publishing.

Thomson, William. "The Injurious Effects of the Air of Large Towns on Animal and Vegetable Life, and Methods Proposed for Securing a Salubrious Air." *Van Nostrand's Eclectic Engineering Magazine,* vol. 20, June 1879.

Thomson, Sir William. *Reflections on the Motive Power of Heat and on Machines Fitted to Develop That Power, from the Original French of N. L. S. Carnot.* New York: John Wiley and Sons, 1890.

Thulesius, Olav. *The Man Who Made the Monitor: A Biography of John Ericsson, Naval Engineer.* Jefferson, NC: McFarland, 2007.

Thurston, R. H. *A History of the Growth of the Steam-Engine.* New York: D. Appleton, 1878.

Tobey, Ronald C. *Technology as Freedom: The New Deal and the Electrical Modernization of the American Home.* Berkeley: University of California Press, 1996.

Torchio, Philip. "Discussion on the Present Limits of Speed and Power of Single Shaft Steam Turbines." *Transactions of the American Institute of Electrical Engineers* 38, no. 2 (1919): 1547–65.

Trachtenberg, Alan. *The Incorporation of America: Culture and Society in the Gilded Age.* 1st ed. New York: Hill and Wang, 1982.

Trollope, Anthony. *North America.* New York: Harper and Brothers, 1862.

Truman, Benjamin Cummings. *History of the World's Fair: Being a Complete and Authentic Description of the Columbian Exposition from Its Inception.* New York: E. B. Treat, 1893.

"Turn the Switch and Heat Your Bath." *Popular Mechanics,* February 1920, 50.

Turner, Frederick Jackson, and John Mack Faragher. *Rereading Frederick Jackson Turner: "The Significance of the Frontier in American History" and Other Essays*. Reprint. New Haven, CT: Yale University Press, 1994.

Twain, Mark. *A Connecticut Yankee in King Arthur's Court*. New York: Harper and Brothers, 1889.

Tyler, John M., John P. Squire, George C. Burpee, and W. O. Robson. *The Official Record of the State Board of Health of Massachusetts Together with a Phonographic Report of the Evidence and Arguments at the Hearing*. Cambridge, MA: Welch, Bigelow, 1874.

Uekötter, Frank. *The Age of Smoke: Environmental Policy in Germany and the United States, 1880–1970*. Pittsburgh, PA: University of Pittsburgh Press, 2009.

Vanden Berg, Michael D. *Annual Review and Forecast of Utah Coal Production and Distribution—2008*. Salt Lake City: Utah Geological Survey, 2010.

Van Dulken, S. *Inventing the 19th Century: 100 Inventions That Shaped the Victorian Age from Aspirin to the Zeppelin*. New York: New York University Press, 2001.

"Ventilation." *Cleveland Medical Gazette* 1, no. 7, January 1, 1860, 189–99.

Visitors' Guide to the Centennial Exhibition and Philadelphia. Edited by the Centennial Board of Finance. Philadelphia: J. B. Lippincott and Company, 1875.

Walbank, W. M. "Lachine Rapids Plant and the Cost of Producing Power for Generating Electricity Therefrom." In *Proceedings of the Twenty-First National Electric Light Association Convention*, ed. National Electric Light Association Convention, 257–76. New York, 1898.

Wallace, Alfred Russell. *The Progress of the Century*. New York: Harper and Brothers, 1901.

"Washing Smoke." *Scribner's Monthly*, January 1876, 453.

Wasik, John F. *The Merchant of Power: Samuel Insull, Thomas Edison, and the Creation of the Modern Metropolis*. 1st ed. New York: Palgrave Macmillan, 2006.

"Watt's What." *National Electric Light Association Bulletin* 7, no. 7 (1920): 521–24.

Weart, Spencer R. *The Discovery of Global Warming*. Cambridge, MA: Harvard University Press, 2003.

Weesner, C. W. *History of Wabash County, Indiana*. Chicago: Lewis Publishing, 1914.

Wermiel, Sara E. "Did the Fire Insurance Industry Help Reduce Urban Fires in the United States in the Nineteenth Century?" In *Flammable Cities: Urban Conflagration and the Making of the Modern World*, edited by Greg Bankoff, Uwe Lübken, Jordan Sand, and Stephen J. Pyne, 235–53. Madsion: University of Wisconsin Press, 2012.

Westinghouse, George, Jr. "A Reply to Mr. Edison." *North American Review* 149, no. 397 (1889): 664.

White, Trumbull, and William Igleheart. *The World's Columbian Exposition, Chicago, 1893.* Boston: J. K. Hastings, 1893.

Whitman, Walt. *Leaves of Grass (1855 First Edition Text).* Radford, VA: Wilder Publications, 2008.

Whitman, Walt. *The Works of Walt Whitman.* Hertfordshire, UK: Wordsworth, 1995.

Whitman, Walt, and G. Schmidgall. *Walt Whitman: Selected Poems 1855–1892.* New York: St. Martin's, 2000.

Whitman, Walt, and M. Warner. *The Portable Walt Whitman.* New York: Penguin Books, 2004.

Wiebe, Robert H. *The Search for Order, 1877–1920.* London: Macmillan, 1967.

Williams, Michael. "Clearing the United States Forests: Pivotal Years 1810–1860." *Journal of Historical Geography* 8, no. 1 (1982): 12–28.

"Wizard Wonders at the Fair." *Omaha Daily Bee,* July 20, 1893.

Worster, Donald. *A River Running West: The Life of John Wesley Powell.* New York: Oxford University Press, 2002.

Worthington, G. "Thirty Years of Electrical Supply in New York." *Electrical Review* 61, no. 11 (1912): 486.

Zangwill, Andrew. *Modern Electrodynamics.* New York: Cambridge University Press, 2013.

INDEX

acid rain, 154
Adams, Edward Dean, 114
Adams, Henry, 4, 102, 188n42; dirty engine house of, 112, 120, 156
Adams, John, 32, 76, 112
advertising, 115, 123, 137, 189n54
Aeschylus, 23
Age of Coal, 123
Age of Electricity, 17, 112, 123, 142, 143
agrarianism, 13, 31, 32, 164n5
agricultural society, 13, 32, 34
air quality, 53, 153, 172n35
Alcott, Amos Bronson, 37
algae bloom, vii
Allis, E. P., 109
Allis-Corliss engines, 108, 109
alternating current, 18, 106, 109, 121, 124; adoption of, 103–4; direct current and, 100, 186n72; voltage of, 99
America by Design (Noble), 136
American Board of Commissioners for Foreign Missions, 71
American Gas Light Journal, 56
American Institute of Electrical Engineers, 110
American Journal of Medical Sciences, 52
American Society of Mechanical Engineers, 58
American Woman's Home: Or, Principles of Domestic Science, The (Beecher and Stowe), 53
Anderson, Benedict, ix
Annual Report of the American Railway Master Mechanics, 57
antismoke activism, 53–54
antismoke ordinances, 49, 54
appliances, 136, 137, 138
arc lighting, 92, 94, 99, 151
Arrhenius, Svante, 55, 152, 173n56
Arthur, King, 117
Atlantic, 162n17

Atlantic Monthly, 67
"Atoms for Peace" speech (Eisenhower), 146
automobiles, electric, 7, 155, 157
Ayanz y Beaumont, Jerónimo de, 42, 167n42

Bacon, Francis, 113
Ballard, Walter J., 122
Baltimore Sun, on Morse telegraph, 91
Banner, Stuart, 187n7
Barrett, John Patrick, 109
Bartlett, Helen: on electricity, 140
Bartlett engine, coal for, 42–43
Basalla, George, 70, 147
Batchelor, Charles, 104
Beecher, Catharine, 22, 23, 24; indoor air quality and, 53; ventilation and, 47
Bell, Alexander Graham, 64, 71, 74, 150; single-pole transmitters and, 72; telephone of, 83, 91
Bellamy, Edward, 24, 29–30, 140, 190n62; electricity and, 20, 21–22; environment and, 22; fire and, 23; utopian society and, 21, 118, 119
Bellamy, Joseph, 37
Bigelow, Jacob, 63, 113
Bismarck Weekly Tribune, 110
Bohn, Frank, 142, 143, 144, 155
boilers, 4, 69, 74, 111
boosterism, 14, 50, 143
Borg, Kevin, 64–65
Bormann, F. Herbert, 153

Bosler, Frank, 19, 132, 151
Boulton, Matthew, 42, 125, 168n42
Boys' Life, 138
Brown, Alfred S., 104, 150, 151
Brown, Harold P., 105
Brown, James M., 97
Browne, Sir Thomas, 26–27, 30
Brush, Charles F., 15, 79, 105, 151; alternating current and, 18; arc lighting system of, 92, 94, 99, 151; dynamos of, 93, 94
Brush Dynamo Electric Machine, 93
Buffalo, electricity for, 114, 115, 119
Burnham, Daniel H., 108

Caesar's Column: A Story of the Twentieth Century (Donnelly), 118
California Electric Light Company, 92–93, 95
Callendar, Guy Stewart, 152
caloric engine, 78, 149, 150
caloric ship, 18
capital punishment, 104–5
capitalism, 14, 20, 33, 63, 136–37, 152, 176n73; carboniferous, 145; coal-based, 119, 121; excesses of, 119
carbon dioxide, 26, 44, 153, 156, 169n55, 174n56; fossil fuels and, 55, 173n56
carbon electrodes, 92
carbon emissions, 50, 70
carbon rods, 92

Carnot, Nicolas-Leonard-Sadi, 148
Carroll, Charles, 76
Carter, Jimmy, 154
Cassier's Magazine, 137
cell phones, 155
"Centennial Exhibition, The" (*New York Times*), 68
"Centennial Outlook, A" (Norton), 77
Chamberlain, Thomas, 173n56
Chandler, Alfred D., Jr., 35, 41
Charles I, King, 25
Charles II, King, 27
Chase, Samuel, 76
Chicago: air pollution in, 53; power for, 121; smoke ordinances in, 54; smokiness in, 44, 55; transformation of, 65
Chicago Citizens' Association, 53
Chicago Daily Tribune, 110, 112
Chicago Department of Health, 54
Chicago Edison Company, 10, 17, 124, 126, 127, 128, 129
Chicago Record, 110
Chicago River, 122
Chicago School of Sanitary Instruction, 140
Chicago Tribune, 52, 108, 109, 122; Fisk Street Station and, 133
Chicago World's Columbian Exposition (1893), 11, 16, 103, 108, 109–10, 112, 121; alternating current and, 106; electrical power and, 107, 126; impact of, 113; power station and, 114
chimneys, 25, 49, 54, 57, 111; height of, 58, 59, 80
Church of England, 25
Cincinnati, 9, 40, 44–45; air pollution in, 53; fuel for, 45; smoke discharge laws in, 49; smokiness of, 44
Citizens' Smoke Abatement Association, 54
City in History, The (Mumford), 9
City of St. Paul v. Gilfallan, The (1886), 49
Civil War, 14, 57, 59
civilization, 9, 31; fire and, 23; neotechnic phase of, 144–45; paleotechnic phase of, 145; power and, 39–40
Clean Air Act (CAA) (1970), 153
Cleveland, 9, 18, 94, 108, 151; air pollution in, 53; Brush system in, 95; coal for, 40; generating system in, 92; smokiness of, 44, 45, 53, 54, 55
Cleveland, Grover, 108
Cleveland Division of Health, 45
Cleveland Plain Dealer, 94
climate change, 155, 157, 173n56, 174n56
coal, ix, 18, 19, 31, 34, 42, 43, 66, 101, 144; anthracite, 40–41, 56, 68, 69, 130; atmospheric decline and, 152; bituminous, 35, 40, 54–55, 56, 130;

burning, 3, 4, 53, 59, 60, 69, 73, 125, 132, 134, 145, 147, 163n2; consumption of, 41, 44, 59, 141, 145, 169n55; cost of, 128, 129; electricity and, 5, 6, 17, 89, 123, 125, 130, 135, 152, 155; emissions from, 157; as energy, 49; environment and, 155; fading of, 81; future without, 149; portability of, 39–40; production of, 44, 68; as regressive, 98; smoke and, 48; smokeless, 134; steam and, 73, 78; transporting, 69
coal dust, 50
coal gas, 52, 55; asphyxiation and, 172n35; property damage from, 53
coal interests, 78–79, 152
Coketown, 10
Columbus, Christopher, 108, 109, 110
Commonwealth Edison, 193n31
communal energy, anticapitalist visions of, 148
communications, 64, 65, 89, 90; business, 91; changes for, 61; electricity and, 4; spatial dynamics of, 82
Connecticut Yankee in King Arthur's Court, A (Twain), 117
consequentiality, 78, 123, 156; environmental, ix, viii, 13, 16, 154, 157
conservation movement, 32
consumer price index, 154

consumerism, 6, 17, 59, 123, 135–36; electrification and, 136, 137, 138; inconsequential, 155–56
consumption, 5, 14, 15, 39, 138, 145; commodity, 65; cultural behaviors of, 7; electrical generation and, 10, 146; first generation of, 157; inconsequential, 6; technology of, 156
Contributions to the Centennial Exhibition (Ericsson), 78–79
Cooke, William Fothergill, 90
Copeland, James, 52
Corliss steam engine, 42, 66, 71, 73, 74, 75, 78, 79, 80, 107, 108, 113, 150; exhibition of, 3; impact of, 68–69; power of, 67; technology and, 70; waterpower for, 43
Cowan, Ruth Swartz, 138
Coxe, Tench, 13, 33
Crockett, Davy, 35–36
Cronon, William, 9; on indigenous peoples, 29; railroad technologies and, 65–66
cryophorus, 151
cultural messages, 8, 137, 145; electricity/energy and, ix; long-standing, ix-x
culture, 5, 7, 156; American, 96, 121, 140, 147; consumer, 6, 20, 123, 135; energy, 11, 17–18, 58, 147; popular, 88

Darwin, Charles: fire discovery and, 23

Dean, Teresa: on White City, 109–10
deforestation, 25, 26, 27
Descartes, René, 87
Dickens, Charles, 10, 36, 38
Dictionary of Practical Medicine, 52
direct current, 97, 104, 126; alternating current and, 100, 186n72
dirty engine house, 112, 120, 156
"Discoveries and Inventions" (Lincoln), 62, 149
distribution networks, 6, 15, 39, 99, 100, 123
Dodd, Anna Bowman, 120
Domesday book, 25
Donnelly, Ignatius, 118, 119
Drexel, Anthony, 107
DuPont de Nemours, P. S., 33–34
Dyer, Frank, 98
dynamos, 93, 94, 95, 102, 109, 111, 116, 121; belt-driven, 132; coal-driven, 78, 79; electricity from, 112; Gramme, 72, 73, 74, 78, 92, 147–48; hydropower-driven, 119, 147–48; Siemen, 147–48; Wallace-Farmer electrical, 71, 72, 73, 74, 78, 92

E. Goddard and Sons Mill Company, 50
Earth Day, 157
"Earth, Water, Aire and Fire" (Higginson), 25
economic development, 13, 46, 59, 117
Edison, Thomas, 11, 22, 24, 79, 93, 101, 103, 107, 110, 111, 123–24, 125, 155; Brush and, 94; dynamos and, 95; electricity and, 91, 96, 98; electrocution and, 105; experimental installation by, 185n68; fire and, 23; first plant of, 122; gaslight and, 96; incandescent lighting and, 99, 100; Insull and, 126; inventions by, 150; Menlo Park and, 74; patents by, 96, 183n40; Pearl Street Station and, 145; personality of, 106; stature of, 96–97; Tesla and, 104; Westinghouse and, 105, 106
Edison Electric Illuminating Company, 97, 105, 151, 185n68; commercialization by, 116; losses for, 106; market share for, 104; "Tower of Light" of, 109; Westinghouse and, 114
"Edison's Newest Marvel" (*New York Sun*), 94
Edward, King, 26, 27
Edwards, Jonathan, 37
Effluent America: Cities, Industry, Energy, and the Environment (Melosi), 9
Eisenhower, Dwight D., 146–47
"Electric Age: A New Utopia, The" (Bohn), 3, 142
electric belts, 139

Electric Building, 110
electric light, 92, 135–36, 140, 183n42
electric motors, 88, 104, 109, 111, 140, 151
Electric Shop, 137
Electric Storage Battery Company, 132
electrical charges, discovery of, 84
electrical current, 82, 93, 125
Electrical Engineer, 106
electrical force, 77, 87, 115; magnetism and, 85, 86
electrical infrastructure, vii, viii, 6, 121, 123, 126; alternate, 148; evolution of, 10; formation of, 16–17
"Electrical Magic" (*Scientific American*), 82
electrical systems, 3, 11, 74, 95
electricity, vii, 72, 80–81, 93, 107, 115, 142, 143, 145; adoption of, 99, 106, 148, 156; advances in, 21; animal, 84, 116; as applied art, 83; benefits of, 144; civilized form of, 9; coal and, 5, 6–7, 17, 89, 116, 123, 125, 130, 135, 152, 155; conceptualization of, 15, 85, 86–87, 152; consumption of, 15, 17, 121, 136, 140, 156; cost of, 154; as curative power, 138, 139, 140; delivery of, 5, 101, 110; demand for, 133, 135; discovery of, 83; energy and, 3–4, 88, 101; environmental consequences and, 16, 157; as exceptional energy source, 121; fire and, ix, 22–23; inconsequentiality of, 16, 156; invisible force of, 16, 74, 82, 116; lightning and, 83; magnetism and, 85, 86, 118; modernity and, 6, 109–10, 112, 123, 155; nature and, 87, 117; as panacea, 103, 186n76; practical applications of, 15, 82; as primary energy source, 111; problems with, 82–83; as progressive, 22, 98; science of, 84–85; societal views of, 121; sources of, 75; as sovereign energy, 81; as stand-alone energy source, viii, 17; static, 84; steam and, 5, 102; substitution of, 89; technology of, 11, 14–15, 73, 77, 82, 83, 103, 117, 147; terms/expressions related to, 189n55; transmission of, 10, 65, 91, 112, 113, 114, 123, 135, 146; utopian narrative of, 16, 77, 145
Electricity Building, 110
Electricity in Health and Disease: A Treatise of Authentic Facts for Readers . . . (Monell), 138–39
Electricity in Medicine: A Practical Exposition of the Methods and Use of Electricity in the Treatment of Disease . . . (Jacoby and Jacoby), 139

Electricity in the 17th and 18th Centuries: A Study of Early Modern Physics (Heilbron), 84
electrification, viii-ix, 10, 13, 23, 24, 123–24, 135–36, 149, 153–54; coal and, 6; consumerism and, 136, 137, 138; development/adoption of, 5; energy exceptionalism and, 7; as energy system, 147; as essential force, 156; fire and, 4; impact of, 17–18; national affairs and, 197n9; nature of, 143; as Progressive technology, 137; rationality/social reform and, 11; social meaning of, 11, 131, 141; statistics/household, 191n3; story of, 8
Electrifying America: Social Meanings Of A New Technology, 1880–1940 (Nye), 10–11
"Electro Magnetic Motor," 104
electro-metallurgy, 82
electrocutions, 104–6
electromagnetism, 85, 86, 87–88, 90, 182n18
electroplating device, 72
electroscope networks, 118
electrostatic induction, theory of, 84
Elements of Technology (Bigelow), 63
Elsworth, Annie: message to, 91
Embargo Act (1807), 33
Emerson, Ralph Waldo, 27, 39, 62, 102, 162n17

Empire District Electric Company, 130
energy: abundance of, x, 5, 13; alternate, 18, 19, 56, 89, 148, 152; clean, 5, 14; coal-based, 18, 31, 43, 45, 123; conceptualization/commoditization of, 42; delivery of, 22; dynamic, 82, 150; electrical, 88, 90, 99, 100, 109, 119, 157; evolution of, 39; exothermic, 23; fire-based, 5, 6, 24, 59; historiography of, 29; inconsequential, x, 5, 156; industrial, 34; mechanical, 42, 114; modernity and, 145; natural, 45, 163n3; perceptions of, 12, 102; portability of, 39–40; public attitudes toward, 153–54; smoke-free, 39; sources of, 4, 20, 75, 81; sustainable, 19, 39, 148; using, 4, 8, 46, 60
energy abstraction, 5, 15, 72, 74, 102, 103, 118, 123
energy consciousness, x, 6, 145, 146, 157
energy consumption, viii, x, 9, 12, 15–16, 43, 45, 59, 101, 118, 140, 145, 147, 155, 177n74; culture of, 17–18; electrification and, ix; energy production and, 5, 16, 23, 31, 58, 66, 78, 82, 111, 125, 157; increase in, 123; limiting, 60; process of, 67; spatial dynamics of, 22; unlimited, 123

energy exceptionalism, ix, 5, 14, 16, 25, 32, 144; electrification and, 7; foundation of, 13, 154; ideology of, 3, 6, 18, 21

energy literacy, x, 6, 156

energy paradigm, electrical power and, 3–4

energy production, viii, 6, 17, 19, 42, 59, 74, 97, 119, 129, 130, 131, 145; coal and, 18, 125, 133, 147, 153–55; decrease in, 123; energy consumption and, 5, 10, 16, 23, 31, 46, 58, 66, 78, 82, 111, 125, 146, 157; fuel for, 12; negative consequences of, 30; planning phase for, 114; process of, 67; salience of, 25; smoke and, 26; spatial dynamics of, 15, 22

energy sources, 21, 33, 39, 41, 60, 73; clean/progressive, 6; development of, 12–13; European, 29; fire as, 23; postcolonial, 152; primary, 31, 39, 43, 111, 148; secondary, viii, 156; unreliable, 132; utopian, 103

Energy Transitions: History, Requirements, Prospects (Smil), 12

engineering, 51, 176n73

Engineering, 99

engines, 111; coal, 6; design of, 125; fire, 167n42; pump, 167n42

environment, 20, 33; coal and, 155; exploitation of, 8, 157; fire and, 22; industrialization and, 8; technology and, 8

environmental consequentiality, viii, 13, 154; cultural discourse of, ix; electricity and, 16, 157

environmental exceptionalism, 45, 46; culture of, 147; ideology of, 21, 52

environmental issues, 8, 13, 28, 37, 68, 77, 78, 147; coal and, 154

Environmental Law and Policy Center, 122

environmental legislation, 26

environmentalism, 8, 9, 10, 47, 177n75

Equality, 118

Ericsson, John, 18, 38, 39, 131, 149, 152, 154; on Centennial Exhibition, 78–79; inventions by, 150; solar power and, 78–80, 151

Ericsson (ship), 149

Ericsson Cycle, 150

Erie Canal, 114

ether, 85, 86, 182n18

Etzler, John Adolphus, 18, 38, 39, 78, 131, 151, 152, 154; energy pursuits of, 148–49; patents and, 166n28

Evans, Lewis: on wealth/power, 32

Evans, Oliver: steam engines and, 42

Evelyn, John, 27, 34

Evershed, Thomas, 114

Everyday Housekeeping, 53
Evolution of Technology, The (Basalla), 147
exceptionalism, 15, 24, 36, 58, 121
executions, 105–6

factories, zoning for, 58
Fairview Park, 69
"Far Worse than Hanging" (*New York Times*), 106
Faraday, Michael, 15, 86–87, 89
Faraday Disk, 86
Farmer's Magazine, 51
Field, Marshall, 52–53
Field Museum, 188n25
fire, 5, 17, 81, 96, 101; civilization and, 23; electricity and, ix, 4, 22–23; energy from, 4, 33; environment and, 22; power and, 22; as regressive, 22; wood and, 28
Fisher, Allan C., Jr., 146, 147
Fisk Street Station, 122, 123, 125, 126, 128, 131, 133, 193n31
Flagg, Samuel, 134
Flint, water crisis in, vii
Flying Railroad Train, 189n55
fossil fuels, 7, 26, 31, 43, 45, 152, 157; carbon dioxide and, 55, 173n56; dependence on, 14; using, 34, 55–56, 59
foul air, demonization of, 26
Frank, Adam: on climate change, 157
Frankenstein, or, The Modern Prometheus (Shelley), 84, 116

Franklin, Benjamin, 61, 89, 93, 94, 101, 155; draft woodstove and, 56; electricity and, 15, 84; logical fluid model of, 87; observations of, 83–84
Frederick, Christine: efficiency and, 195n66
Freese, Barbara: labor issues and, 179n24
Friends' Intelligencer, 76
fuels, 4, 12, 40, 45; acquiring, 23, 24, 26; mineral, 40; shortages of, 25–26, 27, 30; sources of, 28, 60, 157. *See also* fossil fuels
"Fumage," 25
Fumifugium: Or, The Inconvenience of the Aer and Smoake of London Dissipated (Evelyn), 27, 34

gadgets, 63, 64, 67, 137
Galvani, Luigi: experiments by, 84
galvanism, 116
galvanometer, 86
Garland, Hamlin, 110
Garrison, William Lloyd, 185n68
gas stocks, 96, 185n60
gases, burning, 55
gasification, 89
gaslights, 4, 95, 96
gasoline, 154, 192n24
Gaulard-Gibbs distribution system, 100
Gaulard-Gibbs Secondary Generator, 99

General Electric (GE), 73, 126, 128, 131, 132, 151; coal-turbine technology and, 19; losses for, 107; Thomson-Houston and, 107

Generall Historie of Virginia, New England, and the Summer Isles (Smith), 28

generation stations, 16, 132; large-scale, 126, 131; number of, 133

Gerry, E., 40

Gibson, Charles Robert: electricity and, 144

Gilded Age, 17, 21, 47, 119, 127

Gillette, King Camp, 24, 119

Gladden, Washington, 48

Goddard, Calvin, 97

Goddard, E., 50

Godey's Magazine, 53

Good Housekeeping, 137–38

Gowen, Franklin B., 68, 179n24

Graham, Sylvester, 37

Gramme dynamos, 72, 73, 74, 78, 92, 147–48

Grant, Ulysses S., 108

graphophone, 143

Great Barrington, lights for, 100, 103–4

Greeley, Horace, 63

greenhouse gases, 55, 134

Griscom, John Hoskins, 22, 51

Gugliotta, Angela, 53, 170n2

Hamilton, Alexander, 13, 32

Hard Times (Dickens), 10

Hartford Electric Company, 126

Hazard, Samuel: on coal smoke, 49

health issues, 51, 52, 153

heat, 55; electricity and, 4

Heilbron, John L., 84, 85, 182n14

Hennepin, Louis, 28

Henry, John, Jr., 76

Henry, Joseph, 15, 74, 150; electricity and, 88, 89; electromagnetism and, 87–88; Morse and, 90; transmutation and, 73

Henry, Patrick, 75

Henry, William Hirt, 75–76

Higginson, Francis, 25, 26, 28, 29, 30, 31

Hill, James J., 189n55

Hirsh, Richard F., 12, 178n15

Historical Statistics of the United States 1789–1945 (US Census Bureau), 44

history: of energy, 11, 12, 18, 84, 148–49; environmental, 9; intellectual, 10; social, 8, 10–11, 66; of technology, 7, 8, 10–11, 23, 66, 124

Hofstadter, Richard, 13

Holden, L. E., 45

Hoover, Herbert, 145

Hoover Dam, ix

Hopkins, Samuel, 37

"How Long Should a Wife Live?" (advertisement), 137

Howells, William Dean, 29–30, 67, 119, 120

Hughes, Thomas Parke, 10, 11, 124, 163n2

Human Drift, The (Gillette), 119
"Humani ignes" (Pliny the Elder), 23
Humphrey, Edward, 70, 71
Hydriotaphia (Browne), 26–27
hydropower, x, 14, 23, 41, 42, 45, 115, 119, 134; demand for, 133, 146; development of, 129, 156; dominance for, 35; electric energy and, 127, 133, 155; future of, 131; manufacturers of, 39; plants, 127, 132; steam versus, 194n45

Iles, George: on electricity, 197n9
Illinois Supreme Court, 49
illumination, 90; coal gas, 53; oil and, 168n48
"Imagined Communities" (Anderson), ix
independence, 45, 66; energy, 154
Independence Hall, 75
industrialization, 3, 13, 31, 33, 36, 39, 43, 47, 60, 63, 70, 142; British-style, 37; environment and, 8; evils of, 48; fire-based energy and, 5; impact of, 21; market-based, 34; science-based, 137
infrastructure, x, 167n38; electrical, vii, viii, 6, 10, 16–17, 121, 123, 126, 148; energy, 59, 145; hidden, 112
Inquiry into the Principles for a Commercial System for the United States, An (Coxe), 33

Institute of Electrical Engineers, 144
Insull, Samuel, 10, 11–12, 17, 124, 128, 130, 131, 152; Edison and, 126, 137; Electric Shop and, 137; electricity and, 126, 127; Fisk Street Station and, 122; reserves for, 193n31
iron, 34, 41, 43

Jackson Park, 108
Jacoby, George, 139
Jacoby, Ralph, 139
James Steam Mill, 42
Jefferson, Thomas, 8, 61; agrarianism and, 32, 164n5; American utopia and, 33; on DuPont, 33–34; industrialization and, 13, 32; oil/illumination and, 168n48; technology and, 62
John, Richard, 91
Johnson, Thomas, Jr., 76
Joliet, Louis, 28
Jones, Christopher, 60
Journal of Science, 88

Keller, Charles M., 169n49
Kelly, Walt, 157
Kemmler, William, 105–6, 111, 187n7
Klein, Maury, 11, 12

Lachine Rapids installation, 127
Lake Erie, 54; algae bloom in, vii

"Lake Tragedy: The Children Probably Suffocated by Coal-Gas, The" (*Chicago Tribune*), 52
lamps, 96; camphene for, 43; design of, 59
Land of Desire: Merchants, Power and the Rise of a New American Culture (Leech), 135
Lane, Mary Bradley, 119, 120
Lasch, Christopher, 64
Leach, William, 135–36
Leaf (car), 7
Lears, T. J. Jackson, 188n42
Leaves of Grass (Whitman), 62
Leete, Mrs., 21, 22, 118, 140
Lewis, John L., 130
Leyden jars, 84, 90
light, 139; electricity and, 4
lightbulbs, 108; commercialization of, 116
lighting systems, 116, 184n44; arc, 92, 94, 99, 151; incandescent, 92, 99, 100, 106
lightning, electricity and, 83
Likens, Gene, 153
Lincoln, Abraham, 60, 62, 149
locomotion, electricity and, 7
Looking Backward: 2000–1987 (Bellamy), 20, 118, 140
Lowell: air in, 35, 36–37, 38; American exceptionalism and, 36; manufacturing in, 36, 37; as utopian future, 35
Loy, Daniel Oscar: poem by, 110

Macbeth (Shakespeare), 26
Machine in the Garden, The (Marx), 7, 102
machinery, transcendentalism and, 102
Machinery Hall, 66, 67, 69, 70, 71, 75, 78, 91
Mackay, Alexander, 35
magnetism, 88; electrical force and, 85, 86, 118
manufacturing, 13, 37, 39, 43, 61; domestic, 33; power for, 34, 35
Marquette, Jacques, 28
Marshall Field's, electric signage for, 136
Martineau, Harriet, 36, 38
Marx, Karl, 41
Marx, Leo, 7, 24, 64; garden and, 13; mechanization and, 102; organic/inorganic and, 8; technological sublime and, 63
Massachusetts Institute of Technology, 58
Massachusetts Medical Society, 52
Massachusetts State Board of Health, 58
McDonald, Forrest, 12
McLane, Louis, 34
McLane Report, 34
McNeill, John R., 163n3
meat packing industry, 65
mechanization, 8, 33, 61, 64, 77, 87; characterizations of, 102; democratization and, 63;

electrical, 137; steam-powered, 18, 63
medical professionals, 52; electricity and, 138–39
medical science, health problems and, 51
Melosi, Martin V., 5, 9
Melville, Herman, 116, 117
Menlo Park, 74
middle landscape, Jeffersonian notion of, 7, 64
Middle West Utilities, 131
Miller, Joaquin, 67
Miller, Perry, 63
mining, 31, 39, 69, 70
Mizora: A World of Women (Lane), 119
Moby Dick (Melville), 116
modernity, 24, 61–62, 63; electricity and, 6, 109–10, 112, 123, 155; energy and, 145; engineering and, 51; illusionary, 7; progress and, 123; steam engine and, 41
Molly Maguires, 68, 69
Monell, Samuel, 138–39
Monitor (ironclad), 18, 38, 78
Morgan, Hank, 117
Morgan, J. P., 97, 107, 114, 126, 137
Morris, Israel W., 31
Morse, Samuel, 15, 61, 63, 88; code system of, 89, 90; telegraph of, 65, 83, 90, 91
motive power, 23, 41, 60, 69, 69, 89, 90, 92, 122, 144, 148, 149, 151

Muhlhausen Emigration Society, 38
Mumford, Lewis, 9, 144, 145

Nash, Roderick, 102
National Centennial Commemoration, 75
National Conference of Social Work, 140
National Electric Light Association, 127, 138
National Electric Light Association Bulletin, 136
National Geographic, 146
nationalism, 64, 70, 77, 145
natural gas, 55, 92, 126, 192n24
natural resources, 47, 48
nature: electricity and, 87, 117; subjugation of, 115; technology and, 73, 76–77
Nature, 27, 162n17
Nature's Metropolis: Chicago and the Great West (Cronon), 9, 65
Naturphilosophie, 85
Nebraska Committee on Public Utility Information, 136
Networks of Power: Electrification in Western Society, 1880–1930 (Hughes), 10
Neufeldt, Leonard, 62
New England Glass Company, 58
New York City, 130; Brush system in, 95
New York Edison, 126
New York prison system, electrocution and, 105

New York Sun, 94, 95
New York Times, 3, 17, 68, 154; on Edison, 107; executions and, 106
New York Times Magazine, 141
Newcomen, Thomas, 42, 167n42
Newcomen engine, 167n42
Newton, Isaac, 87
Niagara Falls, ix, 107, 114, 117, 119, 122, 142; harnessing power of, 16, 103, 112–13, 115, 116, 118, 121; hydropower and, 115, 127; myth of, 133; natural force of, 113
Niagara Falls Power Company, 114, 115–16
Noble, David, 64, 136, 176n73, 178n13; on electrical/chemical industries, 11
Nollet, Jean Antoine, 90
North American Review, 105, 107
Northern Pacific, 189n55
Norton, Mrs. M. B., 77
nuclear power, viii, 89, 126, 146, 147
Nye, David E., 10–11, 35–36, 64, 164n3

ocean thermal energy conversion (OTEC), 151
Ohio River, 40
oil, ix, 154; burning, 60; crisis, 154; illumination and, 168n48; prices, 193n32
Olney, Charles F., 54
Omaha Daily Bee, 110

One Hundred Years' Progress of the United States, 61
Ørsted, Hans Christian, 15, 87, 94, 101; discoveries by, 85; electrical technology and, 86; electromagnetism and, 182n14
Otis, James, 76
"Our Future Motive Power" (Tesla), 151

Paca, William, 76
Palace of Mechanical Arts, 109
"Paleotechnic Paradise: Coketown" (Mumford), 9
Paradise Within the Reach of All Men, Without Labor, by Powers of Nature and Machinery: An Address to All Intelligent Men (Etzler), 38
Paris exposition (1900), 112
Parson-type seventy-five-kilowatt generator, 126
pastoralism, 13, 14, 24, 32, 33, 34, 58; colonial worldview of, 20; cultural artifacts and, 8
patents, 62, 96–97, 100, 165n26, 166n28, 183n40
Patterson, Walt, 191n6
Peabody, Francis: Chicago Edison and, 129, 130
Peabody Coal Company, 134
Pearl Street Station, 22, 24, 97, 98, 99, 103, 124, 125, 126, 145, 185n66; fire and, 23
Peck, Charles F., 104, 150, 151
Pedro II, Dom, 71, 72

Pedro III, Dom, 91
petroleum, 55, 130, 193n32
Pfaelzer, Jean, 117
Philadelphia Centennial Exhibition (1876), 3, 18, 66, 71, 72, 73, 74, 78, 82, 113, 150, 179n25; arc lighting system and, 92; electrical technology and, 93; electricity and, 75, 103, 115–16; impact of, 14, 100; steam power and, 80; telegraph and, 91; telephone and, 64
photovoltaic cells, viii
Pittsburgh, 9, 21, 33, 100; characterization of, 36; coal and, 34–35, 45; fuel for, 40; smokiness of, 35, 44, 49–50, 53, 54
Pliny the Elder, 23–24
Pogo, 157
pollution, 9; air, 25, 37, 46, 52, 53, 55, 134–35, 152, 153, 154
Pomeroy, S. W., 40
Pomeroy Coal Company, 40
Pope, Ralph, 110
Popular Mechanics, 138
Powder River, 122
power, 74, 97, 133, 139; civilization and, 39–40; coal-derived, 35, 155; fire and, 22; generation of, 129, 132, 135; mechanical purposes and, 38. See also hydropower; motive power; nuclear power; solar power; steam power; wind power

Power Makers, The (Klein), 11
power plants, ix, 4, 114, 126, 152, 193n31; clean, 140; coal-fired, 122, 132, 133, 134, 135, 154; consolidating, 12; dynamos in, 111; smokeless, 134, 135; turbine, 134
power sources, viii, 3, 73; modern/progressive, 4; utopian, 16
"Present Condition, Prospects, Beneficent Work, Needs and Obligations, The" (Humphrey), 70
Progressive Era, 5, 24, 32, 45, 55, 58, 113, 136
Progressivism, 6, 7, 24, 54, 93, 101, 124, 138
Promētheús Pyrphóros, 23
prosmoke agendas, 51
Public Service Company of Northern Illinois, 130
Purchas, Samuel, 28, 29
Purchas His Pilgrimes (Purchas), 28
Puritans, persecution of, 25
pyrotechnologies, 14, 24, 25, 29, 37, 43, 56, 58

railroads, 41, 43, 65–66, 76, 106; smoke issue and, 57, 69; transcontinental, 61, 63, 74
Reagan, Ronald, 154
real estate taxes, 53
Reflections on the Motive Power of Heat (Carnot), 148
Rend, W. P., 50

Republic of the Future, Or, Socialism a Reality, The (Dodd), 120
republicanism, 21, 24, 164n5
Righter, Robert W., 151
Riis, Jacob, 43, 47
Rock Island Daily Argus, 110
Rockefeller, John, 111
Rodney, Lucille, 110
Roebling, John Augustus, 38
Roemer, Kenneth, 117
Romance of Modern Electricity, The (Gibson), 144
Roosevelt, Franklin D., 146
Root, John W., 108
Rosen, Christine Meisner, 55, 173n53
Routes of Power: Energy and Modern America (Jones), 12
Royal Institution, 86
Royal Meteorological Society, 152
Rural Electrification Administration, 146
Rutherford, Janice Williams, 138, 195n66
Rydell, Robert, 107, 108

Sabbatarianism, 71–72
Saint Louis, 49; antismoke activism in, 53–54; smoke discharge laws in, 49
Saltonstall, Leverett, 75
"Save The Forests" (*Friends' Intelligencer*), 76
Savery, Thomas, 42, 167n42
Scientific American, 82, 99

Sears, Roebuck and Company, 139
Sellers, Charles Grier, 163n1
Shakespeare, William, 26
Shelley, Mary, 84, 116, 117, 118
Siemen dynamos, 147–48
Slater, Samuel, 33
smartphones, viii, 157
Smil, Vaclav, 12
Smith, John, 28, 29, 30
smog, 154
smoke, 4, 14, 17, 35, 70, 77, 81, 121, 135; air quality and, 53; asphyxiation and, 172n35; burden of, 24; coal, 8, 22, 44, 48, 49, 50, 51, 52, 53, 55, 56, 57, 83; cultural responses to, 59; elimination of, 58, 133, 134; emissions of, 40, 111; energy production and, 26; impact of, 53, 60–61; nuisance of, 47, 51, 58, 80; as prosperity, 50; railroads and, 57; responses to, 47, 51, 59; sea-coal, 27; washing of, 57
smoke abatement, 46, 55, 59, 171n26, 173n53, 173n55; equipment for, 56, 57
smoke control, 55, 56–57, 80, 170n2; patents for, 165n26
smoke ordinances, 48, 54
Smokeless Combustion of Coal in Boiler Plants, The, 134
smokestacks, 6, 49, 122, 133, 153, 155
Smokestacks and Progressives: Environmentalist, Engineers,

and Air Quality in America, 1881–1951 (Stradling), 8
social activism, 47, 51
social forces, 119, 163n2
Social Gospel movement, 47, 48
social issues, 47, 145, 152, 154, 155
social meaning, 11, 70, 141
social reform, 11, 20, 46
Society for the Promotion of Atmospheric Purity, 54
solar power, 39, 78–80, 151, 154
somatic energy, 32, 45, 163n3
soot, 26, 48, 50, 55, 56, 63, 170n7; impact of, 60–61; removing, 57, 59
soot tax, 53
Sovacool, Benjamin K., 12, 178n15
spark control, 56, 59, 165n26
Stamp Act, 76
Standard Oil Company, 111
steam, 15, 34–35, 55, 66, 70, 80, 81, 121, 144, 168n42; adoption of, 41, 42, 168n43; coal and, 73, 78; electricity and, 5, 102; energy from, 4; hydropower and, 194n45
steam engines, 3, 39, 42, 43, 66, 67, 68–69, 71, 74, 75, 77, 89, 109, 168n43; capability of, 148; coal-fired, 125, 163n2; development of, 167n42; modernity and, 41; reciprocating, 125; thermal efficiency of, 129
steam plants, 128; smokeless, 134
steam turbines, 125, 126, 129, 130, 131, 135, 145, 152; coal-fired, 128, 132; development of, 123; implementation of, 127; power from, 123
steel industry, 34, 43, 50
Steinhart, Eric Charles: theories and, 182n18
Stevens, Francis Putnam, 76
Stewart, Mrs. W. H.: on electrical experience, 143–44
stokers, 56, 134, 135
Stone, Thomas, 76
stoves, 59; coal, 163n2; wood, 56
Stowe, Harriet Beecher, 23, 24, 162n17, 190n62; coal furnaces and, 22; indoor air quality and, 53; ventilation and, 47
Stradling, David, 8, 9, 47, 170n2
Strong, Josiah, 46, 47, 48
Swan, Joseph, 96, 100

Taqī al-Dīn, 42, 167n42
Tariff of 1789: 33
Tarkington, Booth, 50
technical advancements, 7, 8, 37, 46, 62, 64, 74, 119
technical approaches, 14, 56, 57–58
Technics and Civilization (Mumford), 144–45
technological abstraction, 16, 80, 106, 107
technological solutions, 14, 55, 58

technological systems, 65, 66, 103, 123
technology, ix, 4, 10, 33, 37, 59, 62, 63, 64, 72, 92, 146, 155, 156; advanced, 16, 103, 137; affordable, 83; applying, 100, 117, 136–37; autonomous, 74; battery-powered, 74; coal, 7, 19; conceptualization of, 67; confidence in, 61; development of, 11, 12–13, 44, 176n73; electrical, 11, 14–15, 73, 77, 82, 83, 88, 102, 103, 117, 147; energy, 23, 24, 36, 42, 45, 65, 78, 148, 166n26; enthusiasm for, 70, 73; environment and, 8; faith in, 21, 80; fire, 24, 96; gateway, 148; modern, 123; nature and, 73, 76–77; novelty, 100; public face of, 111; railroad, 65–66; smoke, 96; social meaning of, 81; solar, 80; speculative, 114; steam, 42, 69, 71; wire, 80
telegraph, 76, 77, 89, 90, 114; invention of, 102–3
telemachon, 74
telephones, 78, 109
Tennessee Valley Authority (TVA), 146
Tesla (car), 7
Tesla, Nikola, 103–4, 114, 115, 124, 150; Edison and, 104; laboratory of, 117; Peck and, 104; vision of, 112
Tesla Electric Company, 11, 104, 107, 151, 152, 190n59

thermal energy, potential, 125
thermodynamic process, 150
Thomas de la Peña, Carolyn, 139–40
Thomson, Elihu, 73, 79
Thomson-Houston Company, 73, 107
Thomson, Sir William (Lord Kelvin), 71, 72, 112–13
Thoreau, Henry David, 8
Three Mile Island, 147
Thurston, Robert Henry, 41
Tocqueville, Alexis de, 20–21
"Tower of Light," 109
transcendentalism, machinery and, 102
transmission systems, 10, 65, 91, 112, 114, 123, 135
transportation, 31, 41, 64, 65, 69, 121, 144; for energy, 61; steam technology and, 42
Traveler from Altruria: Romance, A (Howells), 29–30, 119
Trollope, Anthony, 36, 38
tuberculosis, 138
turbines: coal-fired, 131; efficiency/speed of, 125; experimental stage for, 126; steam, 123, 125, 126, 127, 128, 129, 130, 131, 132, 135, 145, 152
turbogenerator, 125
Turmoil, The (Tarkington), 50
Twain, Mark, 117
Twin City Rapid Transit Company, 130
Tyndall, John, 55, 152, 173n56

Uekötter, Frank, 170n2
Uncle Tom's Cabin (Stowe), 190n62
Underfeed Stoker Company of America, 134
United States Bureau of Mines, 134
United States Census Bureau, 44
United States Congress, 33, 169n49
United States Department of Agriculture, 129
United States Patent Office, 37–38, 61, 166n28, 169n49
United States Supreme Court, 91
urban-industrial landscape, 20
urbanization, 9, 13, 14, 43, 63
"Use of Solar Heat as a Mechanical Motor-Power, The" (Ericsson), 150
Uses and Abuses of Air: Showing Its Influence in Sustaining Life, and Producing Disease, with Remarks on the Ventilation of Houses (Griscom), 22, 51
utility companies, 137, 141, 144, 152
utopia, 21, 32, 33, 36, 77; electricity for, 121, 123, 126; progressive, 63; technological, 120; visions of, 24–25
utopian literature, 103

Vail, Alfred, 91
Vanderbilt, Mrs. Alva, 97–98
vegetable oils, 43
"Vi-Rex Violet Ray" generator, 138
Volt (car), 7
Volta, Alessandro, 15, 84, 85

Wabash *Plain Dealer,* 93
Walbank, W. M., 127, 128
Walden (Thoreau), 8
Wallace, William, 71, 74, 150
Wallace-Farmer electrical dynamo, 71, 72, 73, 74, 78, 92
Wanamaker's, electric signage for, 136
Washington, George, 76
Washington Times, 143
water, 43; consumption of, 41, 179n26; crisis, vii–viii
water utility networks, 64
waterpower. *See* hydropower
waterwheels, 75
Watt, James, 42, 125, 168n42
"We Recommend Electricity" (*Good Housekeeping*), 137–38
Wednesday Club, 54
West, Julian, 21, 118
West 39th Street plant, 126
Westinghouse, George, 79, 100, 103–4, 107, 112, 124
Westinghouse Air Brake Company, 100
Westinghouse Electric and Manufacturing Company, 11, 100, 107, 128, 134, 151; advertising by, 115, 189n54; alternating current and, 104, 105, 106; Edison and,

105, 106, 114; electrocution and, 104–5; generators by, 126; lights from, 109; Tesla and, 104
Westinghouse Machine Company, 134
whale oil, 43, 118
"What Brush Has Done and Edison Proposes to Do" (*Cleveland Plain Dealer*), 94
Wheatstone, Charles, 90
White City, 11, 16, 108, 111, 112, 116, 117, 121; modernity and, 109–10
White House, solar panels at, 154
Whitman, Walt, 67
Whitney, Eli, 61
Wiebe, Robert, 11
Wilderness and the American Mind (Nash), 102
Williams, Michael, 44
Wind Energy in America (Righter), 151
wind power, 18, 23, 39, 56, 178n15; doubts about, 19, 132; potential of, 149

"Wind Turbines and Invisible Technology: Unarticulated Reasons for Local Opposition to Wind Energy" (Hirsh and Sovacool), 12
Wintour, Sir John, 27
Wisconsin Elevator Company, 45
Wollaston, W. H., 150–51
Women's Health Protective Association of Allegheny County, 53
wood, ix, 25, 27, 29, 43; burning, 44, 60; fire and, 28; shortages of, 26; transporting, 40
Works and Days (Emerson), 62

Yost, Benjamin, 68
"Yost Murder Trial: What Detective M'parlan Knows about the Molly Maguires' Secrets, The" (*New York Times*), 68
"You and the Obedient Atom" (Fisher), 146